Information and Instructions

This shop manual contains several sections each covering a specific group of wheel type tractors. The Tab Index on the preceding page can be used to locate the section pertaining to each group of tractors. Each section contains the necessary specifications and the brief but terse procedural data needed by a mechanic when repairing a tractor on which he has had no previous actual experience.

Within each section, the material is arranged in a systematic order beginning with an index which is followed immediately by a Table of Condensed Service Specifications. These specifications include dimensions, fits, clearances and timing instructions. Next in order of arrangement is the procedures paragraphs.

In the procedures paragraphs, the order of presentation starts with the front axle system and steering and proceeding toward the rear axle. The last paragraphs are devoted to the power take-off and power lift systems. Interspersed where needed are additional tabular specifications pertaining to wear limits, torquing, etc.

HOW TO USE THE INDEX

Suppose you want to know the procedure for R&R (remove and reinstall) of the engine camshaft. Your first step is to look in the index under the main heading of ENGINE until you find the entry "Camshaft." Now read to the right where under the column covering the tractor you are repairing, you will find a number which indicates the beginning paragraph pertaining to the camshaft. To locate this wanted paragraph in the manual, turn the pages until the running index appearing on the top outside corner of each page contains the number you are seeking. In this paragraph you will find the information concerning the removal of the camshaft.

More information available at haynes.com
Phone: 805-498-6703

J H Haynes & Co. Ltd.
Haynes North America, Inc.

ISBN-10: 0-87288-101-6
ISBN-13: 978-0-87288-101-3

Disclaimer

There are risks associated with automotive repairs. The ability to make repairs depends on the individual's skill, experience and proper tools. Individuals should act with due care and acknowledge and assume the risk of performing automotive repairs.

The purpose of this manual is to provide comprehensive, useful and accessible automotive repair information, to help you get the best value from your vehicle. However, this manual is not a substitute for a professional certified technician or mechanic.

This repair manual is produced by a third party and is not associated with an individual vehicle manufacturer. If there is any doubt or discrepancy between this manual and the owner's manual or the factory service manual, please refer to the factory service manual or seek assistance from a professional certified technician or mechanic.

Even though we have prepared this manual with extreme care and every attempt is made to ensure that the information in this manual is correct, neither the publisher nor the author can accept responsibility for loss, damage or injury caused by any errors in, or omissions from, the information given.

SHOP MANUAL

INTERNATIONAL HARVESTER

SERIES A-B-C-Cub-H-M-MTA-4-6-W6TA-9
(For complete model breakdown, see Table A)

This manual covers all models of wheel type tractors listed in column two of Table A. To simplify the presentation of material on subsequent pages, the following system is used:

Where the information applies to a particular model, such as the Super MTA, the WDR9, etc., the complete tractor model designation listed in column two of Table A will be used. If, however, the information applies to all models in a series, the complete series designation will be used. For example: If the designation of "Series 4" is used, the information applies to models O4, OS4, W4 and Super W4. Similarly, "Series 9" includes models W9, WR9, WR9S, WD9, WDR9, Super WD9 and Super WDR9. Refer to column one of Table A.

TABLE A

MODEL IDENTIFICATION DATA

I TRACTOR SERIES	II COMPLETE TRACTOR MODEL DESIGNATION	III VERSIONS BUILT
Series A	A, Super A	1, 2
	AV, Super AV	2
Series B	B, BN	3, 4
Series C	C, Super C	2, 3, 4
Cub	Cub, Cub Lo-Boy	1, 2
Series H	H, Super H	2, 3, 4
	HV, Super HV prior 21621	1
	Super HV after 21620	2
Series 4	O4, OS4, W4, Super W4	1
Series M	M, Super M, Super MLPG, Super MTA	2, 3, 4
	MD, Super MD, Super MTA Diesel	2, 3, 4
	MV, Super MV prior 45371	1
	Super MV after 45370, Super MTA Hi-Clearance	2
	MDV, Super MDV prior 45371	1
	Super MDV after 45370	2
	Super MTA Diesel Hi-Clearance	2
	Super MV LPG prior 45371	1
	Super MV LPG after 45370	2
Series 6	O6, OS6, W6, Super W6, Super W6TA	1
	ODS6, WD6, Super WD6, Super WD6TA	1
	Super W6 LPG	1
Series 9	W9, WR9, WR9S	1
	WD9, WDR9, Super WD9, Super WDR9	1

1. Non-adjustable axle   2. Adjustable axle
3. Dual wheel tricycle   4. Single wheel tricycle

LOCATION OF SERIAL NUMBERS

ENGINE. Engine serial number is stamped on side of crankcase.
TRACTOR. The A, B and C tractor serial number plate is located on seat bracket. The Cub tractor serial number is located on side of steering gear. The H, M, 4 and 6 series tractor serial number plate is located on clutch housing or clutch housing cover. The 9 series tractor serial number plate is located on the fuel tank support.

INDEX (By Starting Paragraph) Except Series MTA & W6TA

★ *See Page 5 for Series MTA & W6TA Index* ★

	Series A	Series B	Series C	Cub	Series H & 4	Non-Diesel Ser. M&6	Diesel Ser. M&6	Non-Diesel Series 9	Diesel Series 9
BELT PULLEY	499	499	499	500	505	505	505	505	505
BRAKES	490	490	494	490	494	494	494	494	494
CARBURETOR (Not LP Gas)	101	101	101	102	102	102	102	102	102
CARBURETOR (LP Gas)						110			
CLUTCH	188	188	188	188	188	188	188	188	188
COOLING SYSTEM									
Fan	168	168	168	168	169	169	169	169	169
Radiator	165	165	165	165	(H)166 (4)167	(M)166 (6)167	(M)166 (6)167	167	167
Water Pump			173		174	174	174	174	174
DIESEL FUEL SYSTEM									
Filters & bleeding							122		122
Injection pump							130		130
Nozzles							126		126
Precombustion chambers							129		129
Quick checks							121		121
DIFFERENTIAL	453	453	455	453	458	458	458	460	460
ENGINE									
Cam followers	46	46	46	46	47	47	47	47	47
Camshaft	69	69	69	70	71	71	71	71	71
Conn. rods & bearings	88	88	88	88	88	88	88	88	88
Crankshaft	90	90	90	90	90	90	90	90	90
Cylinder head	30	30	30	31	32	32	33	34	35
Engine removal	25	25	25	25	(H)26 (4)27	(M)26 (6)27	(M)26 (6)27	28	28
Flywheel	94	94	94	94	94	94	94	94	94
Ignition timing	178	178	178	178	178	178	178	178	178
Injection timing							130		130
Main bearings	90	90	90	90	90	90	90	90	90
Oil pump	96	96	96	97	96	96	96	96	96
Pistons & rings	74	74	74	77	79	79	82	79	82
Piston pins	86	86	86	86	86	86	86	86	86
Piston removal	73	73	73	73	73	73	73	73	73
Rear oil seal	92	92	92	92	92	92	92	92	92
Rocker arms	48	48	48		48A	48A	48B	48A	48B
Timing gear cover	58	58	58	58	(H)59 (4)60	(M)59 (6)60	(M59) (6)60	60	60
Timing gears	62	62	62	62	65	65	65	65	65
Valves & seats	38	38	38	38	38	39	40	39	40
Valve guides & springs	43	43	43	43	43	43	43	43	43
Valve timing	55	55	55	55	55	55	55	55	55
FINAL DRIVE									
Axle shafts	463	463	470	465	475	475	475	482	482
Bull gears	463	463	469	465	474	474	474	482	482
Bull pinions	462	462	468	465	473	473	473	481	481
FRONT SYSTEM									
Axle main member	3		3	3	4	4	4	4	4
Steering knuckles	6		6	6	6	6	6	6	6
Tie rods & drag link	5		5	5	5	5	5	5	5
Tricycle type		1	1		2	2	2		
GOVERNOR	155	155	155	158	161	161	144	161	144
IGNITION AND ELECTRICAL SYSTEM									
Distributor	178	178	178	178	178	178	178	178	178
Magneto	181	181	181	181	181	181	181	181	181
Mounting bracket	--	--	--	--	184	184	184	184	184
L.P.-GAS SYSTEM									
Adjustments						110			
Carburetor						111			
Regulator						112			
"LIFT ALL"									
Power cylinders					526	526	526		
Pump					520	520	520		
POWER TAKE-OFF	499	499	499	500	508	508	508	508	508
REAR AXLE	463	463	470	465	475	475	475	482	482
STARTING SYSTEM (DIESEL)									
Adjustments							150		150
Intake manifold							152		152
STEERING GEAR	7	9	9	12	(H)15 (4)18	(M)15 (6)18	(M)15 (6)18	19	19
"TOUCH CONTROL"									
Adjustment	537		537	537					
Cylinder & valves unit	542		542	542					
Lubrication & bleeding	535		535	535					
Pump	539		539	539					
Troubleshooting	536		536	536					
TRANSMISSION									
Basic procedure	221	221	252	240	261	261	261		
Overhaul	226	226	257	245	268	268	268	286	286

Farmall "Cub"

Farmall "Super A"

Series MTA-W6TA INDEX (By Starting Paragraph)

	Non-Diesel Series Super MTA	Diesel Series Super MTA	Non-Diesel Series Super W6TA	Diesel Series Super W6TA
BELT PULLEY	505	505	505	505
BRAKES	497	497	497	497
CARBURETOR	102	102	102	102
CLUTCH	214A	214A	214A	214A
COOLING SYSTEM				
Fan	169	169	169	169
Radiator	166	166	167	167
Water pump	174	174	174	174
DIESEL FUEL SYSTEM				
Filters and bleeding		122		122
Injection pump		130		130
Nozzles		126		126
Precombustion chambers		129		129
Quick checks		121		121
DIFFERENTIAL	458	458	458	458
ENGINE				
Cam followers	47	47	47	47
Camshaft	71	71	71	71
Connecting rod and bearings	88	88	88	88
Crankshaft	90	90	90	90
Cylinder head	32	33	32	33
Engine removal	26	26	27	27
Flywheel	94	94	94	94
Ignition timing	178	178	178	178
Injection timing		130		130
Main bearings	90	90	90	90
Oil pump	96	96	96	96
Oil relief valve	99	99	99	99
Pistons and rings	79	82	79	82
Piston pins	86	86	86	86
Piston removal	73	73	73	73
Rear oil seal	92	92	92	92
Rocker arms	48	48	48	48
Timing gear cover	59	59	60	60
Timing gears	65	65	65	65
Valves and seats	39	40	39	40
Valve guides and springs	43	43	43	43
Valve rotators	53	53		
Valve timing	55	55	55	55
FINAL DRIVE				
Axle shafts	475	475	475	475
Bevel gears	300	300	300	300
Bull gears	474	474	474	474
Bull pinions	473	473	473	473
FRONT SYSTEM				
Axle main member	4	4	4	4
Steering knuckles	6	6	6	6
Tie rods and drag link	5	5	5	5
Tricycle type	2	2	2	2
GOVERNOR	161	144	161	144
HYDRAULIC LIFT				
Control valves	513	513	514	514
Lubrication	510	510	510	510
Pump	512	512	512	512
System checks	511	511	511	511
PTO Non-Continuous	508	508	508	508
PTO Independent Type				
Drive shaft	509C	509C	509C	509C
Driven shaft	509B	509B	509B	509B
Extension shaft	509A	509A	509A	509A
Overhaul rear unit	508D	508D	508D	508D
Reactor bands	508B	508B	508B	508B
REAR AXLE	475	475	475	475
STEERING GEAR	15	15	18	18
TORQUE AMPLIFIER				
Clutch adjustment	215	215	215	215
Clutch overhaul	216	216	216	216
Planetary unit	217	217	217	217
TRANSMISSION				
Countershaft	281	281	281	281
Driving shaft	277	277	277	277
Main shaft	280	280	280	280
Rear frame cover	275	275	275	275
Reverse idler	282	282	282	282
Shifter rails and forks	276	276	276	276

Farmall "Super C"

Farmall "Super M"

CONDENSED SERVICE DATA

TRACTOR MODELS	A-AV	B-BN	Super A-AV	C	Super C	Cub	H-HV-O4 OS4-W4	Super H-HV-W4	M-MV
GENERAL									
Engine Make	Own	Own	Own	Own	Own	Own	Own	Own	Own
Engine Model	C113	C113	C113	C113	C123	C60	C152	C164	C248
Cylinders	4	4	4	4	4	4	4	4	4
Bore—Inches	3	3	3	3	3⅛	2⅝	3⅜	3½	3⅞
Stroke—Inches	4	4	4	4	4	2¾	4¼	4¼	5¼
Displacement—Cubic Inches	113.1	113.1	113.1	113.1	122.7	59.5	152.1	164	247.7
Compression Ratio (Standard)	6:1	6:1	6:1	6:1	6:1	6.5:1	5.9:1	6.1:1	5.65:1
Compression Ratio	5:1	5:1	5:1				4.5:1		4.5:1
Compression Ratio							4.75:1		4.75:1
Compression Ratio							6.75:1		6.87:1
Pistons Removed From:	Above	Above	Above	Above	Above	Above	Above	Above	Above
Main Bearings, Number of	3	3	3	3	3	3	3	3	3
Main & Rod Bearings Adjustable?	No	No	No	No	No	No	No	No	No
Cylinder Sleeves	Wet	Wet	Wet	Wet	Wet	None	Dry	Dry	Dry
Forward Speeds	4	4	4	4	4	3	5	5	5
Generator & Starter Make	D-R	D-R	D-R	D-R	D-R	D-R	D-R	D-R	D-R
Torque Recommendations				— Inside Back Cover of Manual —					
TUNE-UP									
Firing Order	1-3-4-2	1-3-4-2	1-3-4-2	1-3-4-2	1-3-4-2	1-3-4-2	1-3-4-2	1-3-4-2	1-3-4-2
Valve Tappet Gap	0.014H	0.014H	0.014H	0.014H	0.014H	0.013C	0.017H	0.017H	0.017H
Inlet Valve Seat Angle	45°	45°	45°	45°	45°	45°	45°	45°	45°
Exhaust Seat Angle	45°	45°	45°	45°	45°	45°	45°	45°	45°
Ignition Distributor Make	Own	Own	Own	Own	Own	Own	Own	Own	Own
Ignition Distributor Symbol	A	A	A	A	J	D	A	J	A
Ignition Magneto Make	Own	Own	Own	Own	Own	Own	Own	Own	Own
Ignition Magneto Model	H4	H4	H4	H4	H4	J4	H4	H4	H4
Breaker Gap—Distributor	0.020	0.020	0.020	0.020	0.020	0.020	0.020	0.020	0.020
Breaker Gap—Magneto	0.013	0.013	0.013	0.013	0.013	0.013	0.013	0.013	0.013
Distributor Timing—Retard	TDC	TDC	TDC	TDC	TDC	TDC	TDC	TDC	TDC
Distributor Timing—Full Advance	40°B	40°B	40°B	40°B	30°B	16°B	40°B	30°B	40°B
Magneto Impulse Trip Point	TDC	TDC	TDC	TDC	TDC	TDC	TDC	TDC	TDC
Magneto Lag Angle	35°	35°	35°	35°	35°	13°	35°	35°	35°
Magneto Running Timing	35°B	35°B	35°B	35°B	35°B	13°B	35°B	35°B	35°B
Mark Indicating:									
Magneto Impulse Trips	"DC1-4"	"DC1-4"	"DC1-4"	"DC1-4"	"DC1-4"	Notch		—1st Notch—	
Distributor Retard Timing	"DC1-4"	"DC1-4"	"DC1-4"	"DC1-4"	"DC1-4"	Notch		—1st Notch—	
Mark Location			—Flywheel—					—Fan Pulley—	
Spark Plug Make				— Champion, AC or Auto Lite —					
Model				— Champion 15A; AC 85S Com; or Auto Lite BT8 —					
Electrode Gap	0.025	0.025	0.025	0.025	0.025	0.025	0.025	0.025	0.025
Carburetor Model (I-H)						¾"	D10	1¼"	E12
Model (Marvel-Schebler)	TSX	TSX	TSX	TSX	TSX				
Model (Zenith)			—See paragraph 101—						
Model (Carter)				UT	UT	UT			
Float Setting (Marvel-Schebler)	1/4	1/4	1/4	1/4	1/4				
Float Setting (Zenith)	1 5/32	1 5/32	1 5/32	1 5/32	1 5/32				
Float Setting (I-H)						1 13/32	1 53/64	1 5/16	1 5/16
Engine Low Idle RPM	525	525	525	425	425	500	450	425	425
Engine No Load RPM	1540	1540	1540	1875	1875	1800	1815	1864	1595
Belt Pulley No Load RPM	1272	1272	1272	1549	1549	1487	1121	1121	989

SIZES—CAPACITIES—CLEARANCES

(Clearances in thousandths)

	A-AV	B-BN	Super A-AV	C	Super C	Cub	H-HV-O4 OS4-W4	Super H-HV-W4	M-MV
Crankshaft Journal Diameter	2.1245	2.1245	2.1245	2.1245	2.1245	1.6235	*	2.558	*
Crankpin Diameter	1.7495	1.7495	1.7495	1.7495	1.7495	1.4985	**	2.298	**
Camshaft Journal Diameter, Front (No. 1)	1.8115	1.8115	1.8115	1.8115	1.8115	1.8715	1.931	1.931	2.2435
Journal Diameter, (No. 2)	1.5775	1.5775	1.5775	1.5775	1.5775	1.7465	1.806	1.806	2.1185
Journal Diameter, (No. 3)	1.4995	1.4995	1.4995	1.4995	1.4995	0.8725	1.3685	1.3685	1.8685
Piston Pin Diameter	0.91935	0.91935	0.91935	0.91935	0.91935	0.68765	1.10905	1.10905	1.31265
Valve Stem Diameter	0.341	0.341	0.341	0.341	0.341	0.310	0.341	0.341	0.3715
Top Compression Ring Width	1/8	1/8	1/8	1/8	3/32	3/32	1/8	3/32	1/8
Other Compression Ring Width	1/8	1/8	1/8	1/8	3/32	3/32	5/32	3/32	5/32
Oil Ring Width (Not Kerosene)	1/4	1/4	1/4	1/4	1/4	3/16	1/4	1/4	1/4
Oil Ring Width (Kerosene)	3/16	3/16	3/16	3/16	3/16	No	1/4	1/4	1/4
Main Bearings, Diameter Clearance	1.5-3	1.5-3	1.5-3	1-3.5	1-3.5	2-3.5	2-3	1.1-3.7	*
Rod Bearings, Diameter Clearance	1.5-3	1.5-3	1.5-3	1-3.5	1-3.5	2-3	2-3	1.1-3.7	**
Piston Skirt Clearance	3-4	3-4	3-4	2.2-3	2.2-3	1.6-2.4	3-4	2-4	4-5
Crankshaft End Play	4-8	4-8	4-8	4-8	4-8	4-8	4-8	4-8	4-8
Camshaft Bearing Clearance	2-4	2-4	2-4	2-4	2-4	3-12	1.5-3.5	1.5-3.5	1.5-3.5
Cooling System—Gallons	3 1/4	3 1/4	3 1/4	3 1/4	3 3/4	2 7/16	4 1/4	4 1/8	6
Crankcase Oil—Quarts	5	5	5	5	5	3	6	6	8
Transmission & Differential—Quarts	5	5	4 1/2	19	19	1 3/4	24	24	52
Final Drive, Each—Quarts	1 1/2	1 1/2	1 1/2			7/8	HV 3½	HV 3	MV 3½
Add for PTO and/or BP (Pints)	1	1	1	2	2	1/3			

*See paragraph 90. **See paragraph 88.

CONDENSED SERVICE DATA

Tractor Models	MD-MDV	O6-OS6-W6	ODS6-WD6	Super M-ML-MV-MVL	Super MD-MDV	Super W6-W6L	Super WD6	W9-WR9	WD9-WDR9
GENERAL									
Engine Make	Own	Own	Own	Own	Own	Own	Own	Own	Own
Engine Model	D248	C248	D248	C264	D264	C264	D264	C335	D335
Cylinders	4	4	4	4	4	4	4	4	4
Bore—Inches	3⅞	3⅞	3⅞	4	4	4	4	4.4	4.4
Stroke—Inches	5¼	5¼	5¼	5¼	5¼	5¼	5¼	5.5	5.5
Displacement—Cubic Inches	247.7	247.7	247.7	263.9	263.9	263.9	263.9	334.5	334.5
Compression Ratio (Standard)	14.2:1	5.65:1	14.2:1	5.9:1	16.5:1	5.9:1	16.5:1	5.4:1	14.4
Compression Ratio		4.5:1		5.7:1		5.7:1		4.24:1	
Compression Ratio		4.75:1		4.75:1		4.75:1			
Compression Ratio		6.87:1							
Compression Ratio (LP Gas)				6.75:1		6.75:1			
Pistons Removed From:	Above	Above	Above	Above	Above	Above	Above	Above	Above
Main Bearings, Number of	5	3	5	3	5	3	5	3	5
Main & Rod Bearings Adjustable?	No	No	No	No	No	No	No	No	No
Cylinder Sleeves	Dry	Dry	Dry	Dry	Dry	Dry	Dry	Dry	Dry
Forward Speeds	5	5	5	5	5	5	5	‡5	‡5
Generator & Starter Make	D-R	D-R	D-R	D-R	D-R	D-R	D-R	D-R	D-R
Torque Recommendations				Inside Back Cover of Manual					
TUNE-UP									
Firing Order	1-3-4-2	1-3-4-2	1-3-4-2	1-3-4-2	1-3-4-2	1-3-4-2	1-3-4-2	1-3-4-2	1-3-4-2
Valve Tappet Gap	0.017H	0.017H	0.017H	0.017H	0.017H	0.017H	0.017H	0.017H	0.017H
Starting Valve Lobe Clearance	0.060-0.080		0.060-0.080		0.060-0.080		0.060-0.080		0.060-0.080
Inlet Valve Seat Angle	45°	45°	45°	45°	45°	45°	45°	45°	45°
Exhaust Seat Angle	45°	45°	45°	45°	45°	45°	45°	45°	45°
Ignition Distributor Make	Own	Own	Own	Own	Own	Own	Own	Own	Own
Ignition Distributor Symbol	H	A	H	J	H	J	H	B	H
Ignition Magneto Make	Own	Own	Own	Own	Own	Own	Own	Own	Own
Ignition Magneto Model	H4	H4	H4	H4	H4	H4	H4	H4	H4
Breaker Gap—Distributor	0.020	0.020	0.020	0.020	0.020	0.020	0.020	0.020	0.020
Breaker Gap—Magneto	0.013	0.013	0.013	0.013	0.013	0.013	0.013	0.013	0.013
Distributor Timing—Retard	TDC	TDC	TDC	TDC	TDC	TDC	TDC	TDC	TDC
Distributor Timing—Full Advance	8°B	40°B	8°B	30°B	8°B	30°B	8°B	40°B	8°B
Magneto Impulse Trip Point	6½°ATC	TDC	6½°ATC	TDC	TDC	TDC	TDC	TDC	13°ATC
Magneto Lag Angle	15°	35°	15°	35°	7°	35°	7°	35°	15°
Magneto Running Timing	8½°B	35°B	8½°B	35°B	7°B	35°B	7°B	35°B	2°B
Mark Indicating:									
Magneto Impulse Trips—M, 1st Notch	M	Notch	M	Notch	Notch	Notch	Notch	Notch	M
Distributor Retard Timing				First Notch on Fan Pulley					
Mark Location				Fan Pulley					
Spark Plug Make				Champion, AC or Auto-Lite					
Model For Non-Diesels			Series 9: Ch O Com; AC73 Com; ALTT4. Others: Ch 15A; AC 85S Com; AL BT8						
Model For Starting Diesels				Champion 49; AC 18A; Auto-Lite BT15					
Electrode Gap	0.025	0.025	0.025	0.025	0.025	0.025	0.025	0.025	0.025
Carburetor Model (IH)	F8	E12	F8	E12	F8	E12	F8	E13	F8
Float Setting (IH)	9/32	1 5/16	9/32	1 5/16	9/32	1 5/16	9/32	1 5/16	9/32
Calibration				See Parts Book					
Engine Low Idle RPM	450	425	450	425	500	425	500	425	450
Engine No Load RPM	1610	1595	1610	1600	1580	1600	1580	1650	1665
Belt Pulley No Load RPM	998	989	998	992	980	992	980	778	785

SIZES—CAPACITIES—CLEARANCES

(Clearances in thousandths)

	MD-MDV	O6-OS6-W6	ODS6-WD6	Super M-ML-MV-MVL	Super MD-MDV	Super W6-W6L	Super WD6	W9-WR9	WD9-WDR9
Crankshaft Journal Diameter	3.748	‡‡	3.748	2.808	3.748	2.808	3.748	3.248	4.123
Crankpin Diameter	3.248	‡‡‡	3.248	2.548	3.248	2.548	3.248	‡‡‡	‡‡‡
Camshaft Journal Diameter, Front (No. 1)	2.431	2.2435	2.431	2.2435	2.431	2.2435	2.431	2.306	2.431
Journal Diameter, (No. 2)	2.306	2.1185	2.306	2.1185	2.306	2.1185	2.306	2.181	2.306
Journal Diameter, (No. 3)	2.181	1.8685	2.181	1.8685	2.181	1.8685	2.181	1.8685	2.181
Camshaft Journal Diameter, (No. 4)	1.8685		1.8685		1.8685		1.8685		1.8685
Piston Pin Diameter	1.31265	1.31265	1.31265	1.31265	1.31265	1.31265	1.31265	1.50015	1.50015
Valve Stem Diameter	0.3715	0.3715	0.3715	0.3715	0.3715	0.3715	0.3715	0.402	0.402
Diesel Starting Valve Stem Diameter	0.309		0.309		0.309		0.309		0.312
Top Compression Ring Width	3/32	1/8	3/32	3/32	3/32	3/32	3/32	9/64	1/8
Second Compression Ring Width	1/8	5/32	1/8	3/32	1/8	3/32	1/8	9/64	5/32
Third Compression Ring Width	5/32	5/32	5/32	3/32	5/32	3/32	5/32	9/64	3/16
Oil Ring—Width	1/4	1/4	1/4	1/4	1/4	1/4	1/4	1/4	1/4
Main Bearings, Diameter Clearance	2.7-3.7	‡‡	2.7-3.7	1.1-3.7	1.8-4.8	1.1-3.7	1.8-4.8	2-3	2.7-3.7
Rod Bearings, Diameter Clearance	2.3-3.3	‡‡‡	2.3-3.3	1.1-3.7	1.7-4.7	1.1-3.7	1.7-4.7	‡‡‡	2.5-3.5
Piston Skirt Clearance	5.5-6.5	4-5	5.5-6.5	2.9-3.7	4.6-5.4	2.9-3.7	4.6-5.4	6-7	6.5-7.5
Crankshaft End Play	4-8	4-8	4-8	4-8	4-8	4-8	4-8	4-8	4-8
Camshaft Bearing Clearance	1.5-3.5	1.5-3.5	1.5-3.5	1.5-3.5	1.5-3.5	1.5-3.5	1.5-3.5	1.5-3.5	1.5-3.5
Cooling System—Gallons	7	6	7	6¼	7	6¼	7	10	10
Crankcase Oil—Quarts	9	8	9	8	9	8	9	11	11
Transmission & Differential—Quarts	52	52	52	52	52	52	52	40	40
Final Drive, Each—Quarts	MDV3½			MV-MVL3 MDV3					

‡Fifth speed locked out when using steel wheels.   ‡‡See paragraph 90.   ‡‡‡See paragraph 88.

CONDENSED SERVICE DATA

Tractor Models	Non-Diesel Series Super MTA	Diesel Series Super MTA	Non-Diesel Series Super W6TA	Diesel Series Super W6TA	WR9S	Super WD9 & WDR9
GENERAL						
Engine Make	Own	Own	Own	Own	Own	Own
Engine Model	C264	D264	C264	D264	C335	D350
Cylinders	4	4	4	4	4	4
Bore—Inches	4	4	4	4	4.4	4.5
Stroke—Inches	5¼	5¼	5¼	5¼	5.5	5.5
Displacement—Cubic Inches	263.9	263.9	263.9	263.9	335	350
Compression Ratio (Standard)	5.9:1	16.5:1	5.9:1	16.5:1	5.4:1	15.6:1
Pistons Removed From:	Above	Above	Above	Above	Above	Above
Main Bearings, Number of	3	5	3	5	3	5
Main Bearings Adjustable?	No	No	No	No	No	No
Rod Bearings Adjustable?	No	No	No	No	No	No
Cylinder Sleeves	Dry	Dry	Dry	Dry	Dry	Dry
Forward Speeds	10	10	10	10	5	5
Generator & Starter Make	D-R	D-R	D-R	D-R	D-R	D-R
Torque Recommendations			Inside Back Cover of Manual			
TUNE-UP						
Firing Order	1-3-4-2	1-3-4-2	1-3-4-2	1-3-4-2	1-3-4-2	1-3-4-2
Valve Tappet Gap	0.017H	0.017H	0.017H	0.017H	0.017H	0.017H
Starting Valve Lobe Clearance		0.060-0.080		0.060-0.080		0.060-0.080
Inlet Valve Seat Angle	45°	45°	45°	45°	45°	45°
Exhaust Valve Seat Angle	45°	45°	45°	45°	45°	45°
Ignition Distributor Make	Own	Own	Own	Own	Own	Own
Ignition Distributor Symbol	J	H	J	H	B	H
Ignition Magneto Make	Own	Own	Own	Own	Own	Own
Ignition Magneto Model	H4	H4	H4	H4	H4	H4
Breaker Gap—Distributor	0.020	0.020	0.020	0.020	0.020	0.020
Breaker Gap—Magneto	0.013	0.013	0.013	0.013	0.013	0.013
Distributor Timing—Retard	TDC	TDC	TDC	TDC	TDC	TDC
Distributor Timing—Full Advance	30°B	8°B	30°B	8°B	40°B	8°B
Magneto Impulse Trip Point	TDC	TDC	TDC	TDC	TDC	13°ATC
Magneto Lag Angle	35°	7°	35°	7°	35°	15°
Magneto Running Timing	35°	7°	35°	7°	35°	2°
Pulley Mark Indicating Distributor Retard Timing			Last Notch on Crankshaft Pulley			
Spark Plug Size	18mm	18mm	18mm	18mm	18mm	18mm
Electrode Gap	0.025	0.025	0.025	0.025	0.023	0.023
Carburetor Make	Own	Own	Own	Own	Own	Own
Model	E12	F8	E12	F8	E13	F8
Float Setting	1 5/16	9/32	1 5/16	9/32	1 5/16	9/32
Calibration			See Parts Book			
Engine Low Idle RPM	425	500	425	500	425	500
Engine High Idle RPM	1600	1580	1600	1580	1650	1635
Belt Pulley No Load RPM	988	976	988	976	778	771
SIZES—CAPACITIES—CLEARANCES (Clearances in thousandths)						
Crankshaft Journal Diameter	2.808	3.748	2.808	3.748	3.248	4.123
Crankpin Diameter	2.548	3.248	2.548	3.248	2.997	2.748
Camshaft Journal Diameter:						
No. 1 (Front)	2.2435	2.431	2.2435	2.431	2.306	2.431
No. 2	2.1185	2.306	2.1185	2.306	2.181	2.306
No. 3	1.8685	2.181	1.8685	2.181	1.8685	2.181
No. 4		1.8685		1.8685		1.8685
Piston Pin Diameter	1.31265	1.31265	1.31265	1.31265	1.5001	1.5001
Valve Stem Diameter	0.3715	0.3715	0.3715	0.3715	0.402	0.402
Diesel Starting Valve Stem Diameter		0.309		0.309		0.312
Top Compression Ring Width	3/32	3/32	3/32	3/32	5/32	7/32
Second Compression Ring Width	3/32	1/8	3/32	1/8	5/32	7/32
Third Compression Ring Width	3/32	5/32	3/32	5/32	5/32	11/64
Oil Ring Width	¼	¼	¼	¼	¼	11/64
Main Bearings, Diameter Clearance	1.1-3.7	1.8-4.8	1.1-3.7	1.8-4.8	1.9-4.9	2-5
Rod Bearings, Diameter Clearance	1.1-3.7	1.7-4.7	1.1-3.7	1.7-4.7	2-5	1.9-4.9
Piston Skirt Clearance	2.9-3.7	4.6-5.4	2.9-3.7	4.6-5.4	5-6	6.5-7.5
Crankshaft End Play	4-8	4-8	4-8	4-8	6-10	4-8
Camshaft Bearing, Diameter Clearance	1.5-3.5	1.5-3.5	1.5-3.5	1.5-3.5	1.5-3.5	1.5-3.5
Cooling System—Gallons	6¼	7	6¼	7	8½	9½
Crankcase Oil—Quarts	8	9	8	9	11	11
Transmission & Differential—Quarts	60	60	60	60	40	40
Final Drive, Each—Quarts	MTAV3	MTADV3				
Lift-All—Quarts	7	7	8	8		
Live PTO Housing (Rear Unit)—Quarts	2	2	2	2		
TIGHTENING TORQUES—FT.-LBS.						
Cylinder Head	110-115	110-135	110-115	110-135	160	160
Rod Bolts	50-55	110-120	50-55	110-120	87	55
Main Bearing Bolts						
Center	100-105	250-275	100-105	250-275	125	250
Others	100-105	150-175	100-105	150-175	125	150
Torque Recommendations			Inside Back Cover of Manual			

FRONT SYSTEM—TRICYCLE TYPE

Series B-C

1. Dual front wheels are mounted on a horizontal axle which is riveted to the lower portion of the bolster as shown in Fig. IH300. The horizontal axle can be separated from the bolster by removing the four rivets. Replacement bolster or horizontal axle are furnished with new rivets; or, the bolster and horizontal axle are available as a pre-riveted assembly. The single front wheel is mounted in a wheel fork as shown in Fig. IH301. In either case, the wheel fork or bolster is retained to the lower portion of the steering worm shaft by four stud nuts. The procedure for removing either the wheel fork or bol-

ster is evident after an examination of the unit.

Series H-M

2. The single front wheel is mounted in a fork and double front wheels on a lower bolster and axle combination. Refer to Figs. IH300 and 301. In either case, the fork or lower bolster is bolted to the lower portion of the upper bolster pivot shaft. Procedure for removal of the fork or lower bolster is evident after an examination of the unit. The single front wheel on some very late models is equipped with taper roller bearings and the construction differs from that shown in Fig. IH301.

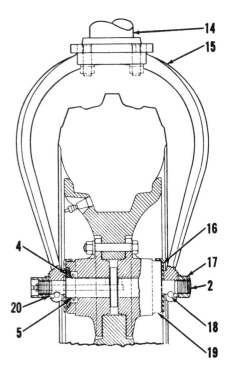

Fig. IH301—Sectional view of a typical fork mounted single front wheel on the B, BN, C and Super C tractors. Series H and M are similar except for some late models which are equipped with taper roller bearings.

2. Front axle	16. Hub shield
4. Felt washer	17. Axle nut
5. Axle oil seal	18. Retaining bolt
14. Worm wheel shaft	19. Hub
15. Wheel fork	20. Nut lock

1. Bolster
2. Axle
3. Dust deflector
4. Felt washer
5. Oil seal
6 & 8. Inner bearing
7. Oil seal ring
9. Front wheel
10 & 11. Outer bearing
12. Gasket
13. Hub cap

Fig. IH300—Exploded view of B, BN, C and Super C bolster, wheel and axle assembly. The bolster and/or horizontal axle are available as individual units, or as pre-riveted assembly. Series H and M are similarly constructed.

FRONT SYSTEM—AXLE TYPE

AXLE MAIN MEMBER

Series A-C-Cub

3. The A, Super A and Cub axle main member pivots on shaft (26—Fig. IH302) which is retained in the axle mounting bracket (steering gear housing base) by clamp bolts. The AV and Super AV axle main member pivots on a shaft which passes through the steering gear housing base and extends rearward where the shaft is anchored to the clutch housing. The two pre-sized axle pivot shaft bushings (21) which are pressed into the axle main member can be renewed after removing the axle main member from tractor.

The C and Super C front axle main member pivots on shaft (26) which is retained in the axle mounting bracket (41) by pin (40). Three pre-sized bushings (21) which are pressed into the axle main member and integral stay rod, can be renewed after removing the axle main member from tractor.

Series H-M-4-6-9

4. On adjustable axle models shown in Figs. IH304, 305A and 305B, the center member pivots on pin (21) which is pinned in the adapter or lower bolster (19). Two pivot pin bushings (12) which are passed into the center member, should be reamed

after installation to provide a pin to bushing clearance of 0.003-0.006.

On non-adjustable axle models, the unbushed main member pivots on pin (24—Fig. IH305) or (10—Fig. IH306). Excessive clearance between axle and pivot pin is corrected by renewing the worn part.

TIE RODS AND DRAG LINK

All Models So Equipped

5. Procedure for removing the tie rod or rods, tie rod tubes, tie rod ends and/or drag link is self-evident after an examination of the unit and reference to Figs. IH302, 304, 305, 305A, 305B or 306.

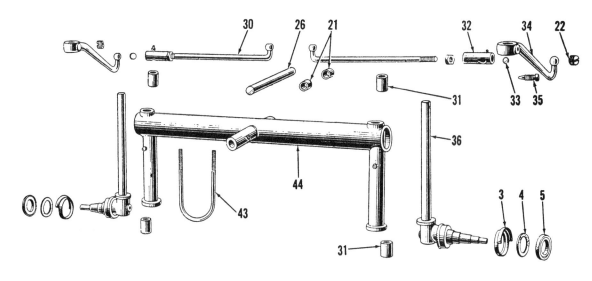

A- super A

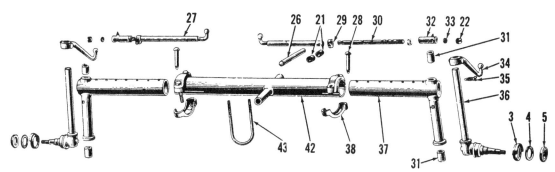

A- super A

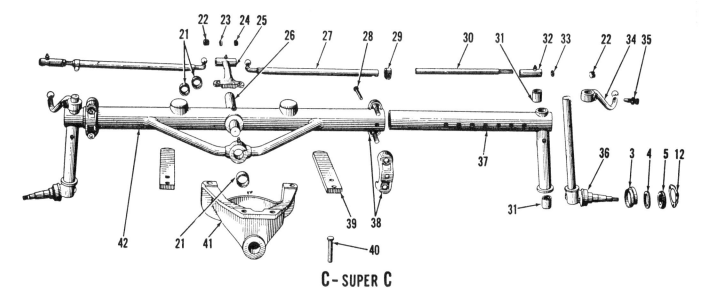

C- super C

Fig. IH302—Exploded views of the axle assemblies which are used on the A, Super A, C and Super C tractors. The adjustable front axle assembly which is used on the Cub, AV and Super AV tractors is similar to the A and Super A assembly shown in the center view. The Cub non-adjustable front axle is similar to the top view. Items (3 & 43) are not used on the Cub; however, axle pivot shaft thrust washers (not shown) are used.

3. Dust deflector	23. Ball center seat		34. Knuckle arm	40. Retainer pin
4. Felt washer	24. Ball end seat	29. Tie rod clamp	35. Knuckle arm set	41. Axle mounting
5. Axle oil seal	25. Center steering	30. Tie rod	screw	bracket
12. Hub cap gasket	arm	31. Knuckle post	36. Knuckle	42. Axle center mem-
21. Pivot shaft bush-	26. Pivot shaft	bushing	37. Axle extension	ber
ings	27. Tie rod tube	32. Tie rod end	38. Extension clamp	43. U-Bolt
22. Ball adjusting seat	28. Clamp pin	33. Ball seat	39. Bumper stop	44. Axle

Adjust the toe-in of the front wheels to approximately ⅛-⅜ inch by varying the length of the tie rod (or rods).

STEERING KNUCKLES

All Models So Equipped

6. Procedure for removing knuckles from axle or axle extensions is evident after an examination of the unit and reference to Fig. IH302, 304, 305, 305A, 305B or 306.

On the A, AV, Super A & AV, C, Super C and Cub, the knuckle post bushings are pre-sized to provide a clearance of 0.002-0.004 (0.001-0.006 on Cub) for the knuckle post.

On other models, the steering knuckle bushings require final sizing after installation. Ream or hone the bushings to provide suggested minimum clearance of 0.003 for the knuckles

Note: On some models, a wheel bearing dust deflector is sweated to knuckle in production; however, the deflector can be satisfactorily installed in the field, without the use of solder, by heating to high enough temperature to facilitate installing same on knuckle.

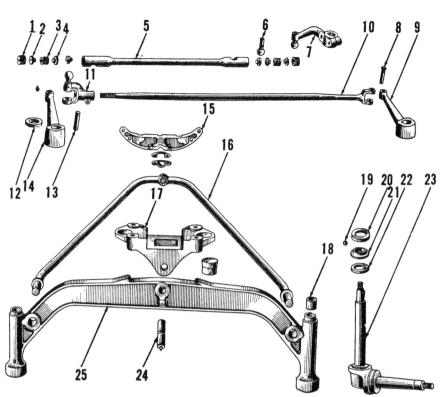

Fig. IH305—Typical exploded view of front axle, mounting bolster and steering arm connections as used on early production high clearance models of the H and M series. For later construction, refer to Fig. IH305A.

1 & 2. Ball nut & seat	8. Clevis pin	14. R. H. steering arm	20. Bearing outer race
3. Spring	9. L. H. steering arm	15. Ball socket	21. Bearing inner race
4. Plug	10. Tie rod	16. Stay rod	22. Felt washer
5. Drag link	11. Tie rod clevis	17. Lower bolster	23. Knuckle
6. Steering arm ball	12. Retaining washer	18. Knuckle bushing	24. Pivot pin
7. Center steering arm	13. Clevis pin	19. Bearing balls	

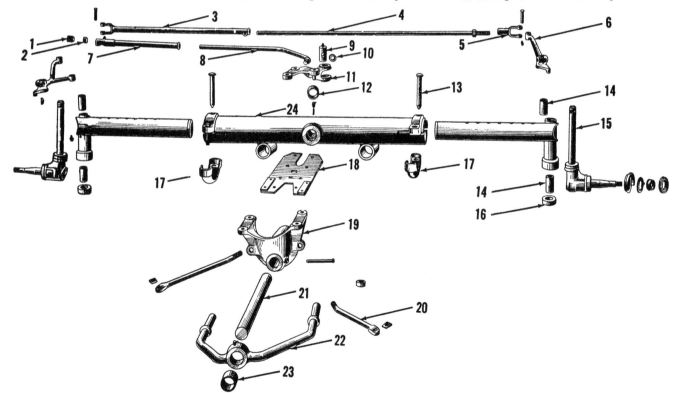

Fig. IH304—Exploded view of front axle and mounting bracket assembly as used on early production series H and M. Pivot pin bushings (12 & 23) require final sizing after installation. For later construction, refer to Fig. IH305B.

1 & 2. Ball seat & nut	7. Drag link sleeve	16. Thrust bearing	21. Pivot pin
3. Tie rod sleeve	8. Drag link	17. Clamp	22. Stay rod
4. Tie rod	9. Steering arm pin	18. Adapter plate	23. Bushing
5. Yoke	10. Washer	19. Adapter	24. Axle center member
6. Steering arm	11. Center arm	20. Brace	
	12. Bushing		
	13. Clamp pin		
	14. Knuckle bushing		
	15. Knuckle		

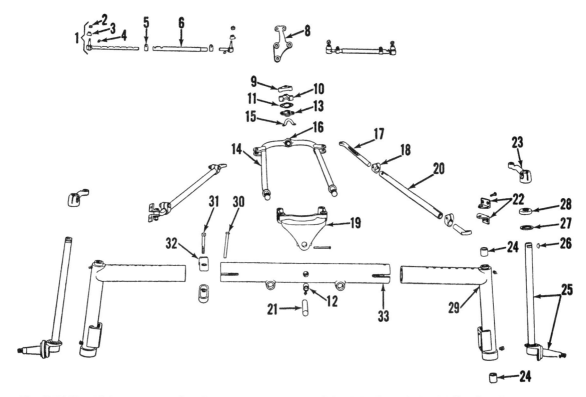

Fig. IH305A—High clearance adjustable axle used on late models of the Super HV and MV series. See legend for Fig. IH305B.

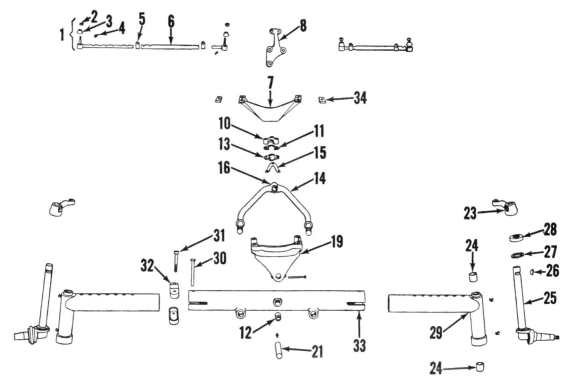

Fig. IH305B—Exploded view of the adjustable front axle used on late models of the Super H and M series. See Fig. IH305A for the late model high clearance axle.

1. Tie rod assembly	7. Ball socket support	14. Stay rod	24. Bushings	29. Axle extension
2. Nut	8. Steering gear arm	15. Lock plate	25. Steering knuckle	30. Clamp pin
3. Dust cover	10. Ball socket	16. Stay rod ball	26. Woodruff key	31. Clamp bolt
4. Fitting	11. Shim	19. Lower bolster	27. Felt washer	32. Axle clamp
5. Clamp	12. Bushings	21. Axle pivot pin	28. Thrust bearing	33. Axle center member
6. Tie-rod tube	13. Ball socket cap	23. Steering knuckle arm		34. Washer

STEERING GEAR

Series A

The non-adjustable steering gear is located within the steering gear housing (45—Fig. IH307) which is bolted to the front face of the engine. The front axle support (steering gear housing base) is bolted to the lower portion of the steering gear housing. Steering arm (25) is retained to the worm wheel shaft (53) by a clamp bolt.

7. REMOVE AND REINSTALL. To remove the steering gear housing, front axle support (steering gear housing base), axle and wheels as an assembly, proceed as follows: Remove hood, grille and radiator, and jack up front of tractor to remove weight from front wheels. To prevent steering gear housing from tilting when housing is disconnected from engine, place wood wedges between axle and steering gear housing base. Remove generator, regulator and mounting bracket from engine. Remove the steering worm shaft. Remove the bolts retaining the steering gear housing to the engine and move assembly away from tractor.

Reinstall the unit by reversing the removal procedure.

8. OVERHAUL. The steering gear unit can be overhauled without removing the assembly from tractor. Remove steering wheel and bearing retainer (57—Fig. IH307). Rotate worm and shaft (59) forward and out of housing (45). The pre-sized worm bushing (58) and seal (60) can be renewed at this time. Install seal with lip of same facing inward toward gears. Use tin sleeve or shim stock when installing the worm shaft to avoid damaging the seal. To remove sector (49) and shaft (53), remove axle assembly from tractor and disconnect steering arm (25) from the shaft. Remove the steering gear housing base retaining cap screws and remove the base. Pre-sized bushing (50) and spring loaded seal (51) can be renewed at this time. Install seal with lip of same facing the gears. Remove the three bearing retainer cap screws (46) and pull sector, shaft and bearing assembly from the housing. To separate the assembly, remove snap ring (47).

Reassemble the gear unit by reversing the disassembly procedure and use tin sleeve or shim stock when

assembling the steering gear housing base to the steering gear housing to avoid damaging seal (51).

Series B-C

The steering worm and worm wheel are located in the steering gear housing, which is bolted to the front face of the engine. The bolster or wheel fork is retained directly to the lower portion of the steering worm wheel shaft by four stud nuts. On the C and Super C adjustable axle models, the center steering arm is bolted to the lower portion of the worm wheel shaft, and the axle mounting bracket is retained to the steering gear housing by four cap screws. The unit is non-adjustable; however, excessive backlash between the worm and worm wheel can be partially corrected by changing

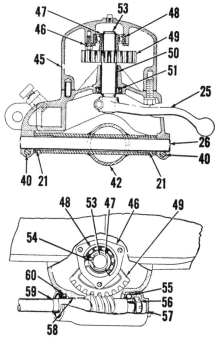

Fig. IH307—Top and side sectional views of the A, AV, Super A and AV steering gear unit. Bushings (50 & 58) are presized.

21. Pivot pin bushings	49. Sector
25. Center arm	50. Bushing
26. Pivot shaft	51. Oil seal
40. Retainer pin	53. Sector shaft
42. Axle	54. Woodruff key
45. Steering gear housing	55. Worm bearing
46. Bearing retainer plate cap screws	56. Worm nut
47. Snap ring	57. Worm bearing retainer
48. Bearing	58. Worm bushing
	59. Worm
	60. Worm oil seal

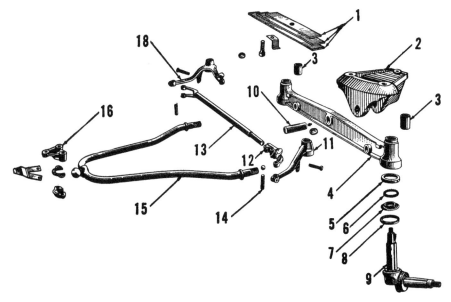

Fig. IH306—Exploded view of series 4 and 6 front axle and associated parts. Series 9 front axle assembly is similarly constructed.

1. Spring	7. Bearing inner race
2. Bolster	8. Felt washer
3. Knuckle bushing	9. Knuckle
4. Axle	10. Pivot pin
5. Bearing outer race	11. R. H. steering arm
6. Bearing balls	12. Tie rod clevis
	13. Tie rod
	14. Clevis pin
	15. Stay rod
	16. Ball socket
	18. L. H. steering arm

the position of the worm wheel on the **worm wheel shaft spline so as to bring** *unworn teeth of the worm wheel into mesh with the steering worm.*

9. **REMOVE AND REINSTALL.** To remove the steering gear housing and front axle and support, bolster or wheel fork as an assembly, proceed as follows: Remove hood, grille and radiator. Remove generator, regulator and mounting bracket assembly from engine. Disconnect the steering shaft front universal joint, and pry universal joint rearward and off the steering worm shaft. Remove radiator drain cap and pipe. Place jack under torque tube (clutch housing) and remove weight from front wheels. Support steering gear housing and remove housing to engine retaining bolts. Jack up tractor high enough for crankshaft pulley to clear steering gear housing and move assembly away from tractor.

10. **OVERHAUL.** The steering gear unit can be overhauled without removing the assembly from tractor.

To remove the steering worm, proceed as follows: Remove grille and drain steering gear housing. Disconnect the steering shaft front universal joint, and slide universal joint rear-ward and off the steering worm shaft. Remove Woodruff key from worm shaft. Remove both starting crank bracket front retaining cap screws and block-up between bracket and steering gear housing enough to permit worm to come out. Remove steering worm bearing retainer (57 — Fig. IH308), and turn worm forward and out of housing. Worm bushing (58) and worm shaft oil seal (60) can be renewed at this time. Install worm shaft oil seal with lip of seal facing inward toward steering gears. Use a tin sleeve or shim stock when reinstalling worm shaft to prevent damaging the seal. To remove worm shaft ball bearing, remove cotter key and nut, and bump worm shaft out of bearing.

To remove the steering worm wheel and shaft assembly, proceed as follows: Remove grille and drain steering gear housing. Jack up tractor under torque tube (clutch housing) and remove bolster or wheel fork on tricycle type models; or, on the C and Super C adjustable axle versions, disconnect the center steering arm from worm wheel shaft, remove four axle mounting bracket to steering gear housing retaining cap screws and move axle and wheels assembly away from tractor. Remove the four worm wheel shaft lower bearing cage and cover retaining cap screws, and bump entire assembly out of steering gear housing. The unit is shown removed in Fig. IH309. To disassemble the unit, remove the worm wheel shaft upper bearing stud nut and pull out the stud (64); place assembly on a suitable press, and press bearings and worm wheel off of worm wheel shaft. At this time, worm wheel shaft lower bearing and oil seal (51—Fig. IH308) can be renewed. Install oil seal with lip of seal facing up toward worm wheel. Use a tin sleeve or shim stock when reinstalling worm wheel shaft to prevent damaging the seal.

Reassemble the unit by reversing the disassembly procedure.

Cub

The worm and worm wheel are located in the steering gear housing (22—Fig. IH310) which is bolted to the front of the engine. The front axle support (steering gear base) (2) is bolted to the lower side of the steering gear housing. The steering gear housing also forms the lower tank for the radiator. The steering gear is non-adjustable, although the worm wheel can be relocated on the shaft splines to bring unworn teeth of the worm wheel into mesh with the steering worm.

12. **REMOVE AND REINSTALL.** To remove the steering gear housing (22) and base (2) (front axle support) and axle as a single unit, proceed as follows: Remove hood and grille and disconnect engine water inlet hose.

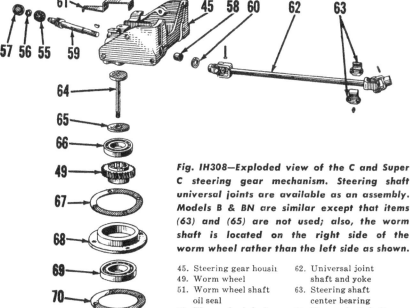

Fig. IH308—Exploded view of the C and Super C steering gear mechanism. Steering shaft universal joints are available as an assembly. Models B & BN are similar except that items (63) and (65) are not used; also, the worm shaft is located on the right side of the worm wheel rather than the left side as shown.

45. Steering gear housing	62. Universal joint shaft and yoke
49. Worm wheel	63. Steering shaft center bearing
51. Worm wheel shaft oil seal	64. Worm wheel shaft upper bearing stud
53. Worm wheel shaft	
55. Bearing	65. **Worm shaft seal**
56. Worm nut	66 & 69. **Bearing**
57. Bearing retainer	67. Gasket
58. Worm bushing	68. Bearing cage
59. Worm	70. **Bearing cage cover gasket**
60. Oil seal	71. **Bearing cage cover**
61. Starting crank bracket	

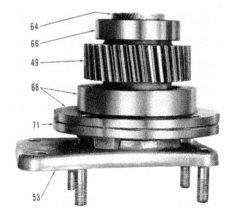

Fig. IH309—Models C and Super C steering worm wheel and shaft assembly removed from the steering gear housing. Models B and BN are similar. The unit can be disassembled after removing stud (64). See legend for Fig. IH308.

Place a support under the engine so that there is little or no load on the front wheels. To prevent steering gear base and housing from tilting, insert on each side a 1 x 2 inch wood wedge between axle and steering gear base. Remove worm shaft bearing (18) and withdraw shaft by rotating same out of mesh with worm wheel. Support steering gear housing (22), remove housing to engine retaining bolts and move assembly forward and away from tractor.

13. **OVERHAUL.** Steering gear unit can be disassembled and overhauled without removing steering gear housing from tractor. To remove steering worm and shaft, remove cap screws retaining steering shaft support to clutch housing. Remove cap screws retaining steering worm shaft bearing (18) to the steering gear housing and withdraw worm and shaft by rotating same out of mesh with the worm wheel.

Removal of worm wheel and shaft is accomplished by first removing front axle from steering gear housing base (2). Remove the center steering arm (8) and Woodruff key (14). Remove cap screws retaining steering gear base to housing and bump the

housings apart. Withdraw steering worm wheel and shaft. Both the upper and lower steering worm wheel shaft bushings (1) and (15) can be renewed at this time.

Series H-M

The non-adjustable worm and sector type steering gear is contained in the upper portion of the upper bolster.

15. **REMOVE AND REINSTALL.** To remove the steering gear, bolster, front wheels, grille and radiator as an assembly, proceed as follows: Remove hood, disconnect the steering shaft universal joint and remove the U-joint Woodruff key. Disconnect radiator hoses and remove dust pan from under front of frame rails. Remove the radiator upper support rod, then support tractor under clutch housing and remove cap screws which retain the upper bolster to the side rails. Using a pry bar, carefully work the front end unit forward and the steering worm shaft out of the steering shaft center bearing. Remove front end unit from tractor.

16. **OVERHAUL.** The steering gear unit can be overhauled without removing the gear unit from the tractor. To remove the steering worm and shaft, remove hood and grille, dis-

connect the steering shaft universal joint and remove the U-joint Woodruff key. Remove the worm shaft bearing retainer (2—Fig. IH312) and withdraw shaft by rotating same forward and out of the steering gear housing. Steering worm shaft bushing (14) which is pre-sized, and oil seal (15) can be renewed at this time.

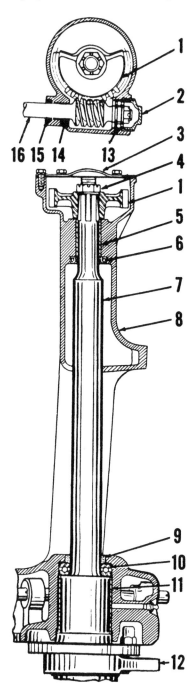

Fig. IH312—Sectional views of series H upper bolster and steering gear assembly. Series M construction is similar.

1. Sector	8. Upper bolster
2. Worm bearing retainer	9. Felt seal
	10. Thrust bearing
3. Gear housing cover	11. Bushing
	12. Lower bolster,
4. Sector nut	fork or arm
5. Bushing	13. Bearing
6. Oil seal	14. Bushing
7. Upper bolster pivot shaft	15. Felt seal
	16. **Worm and shaft**

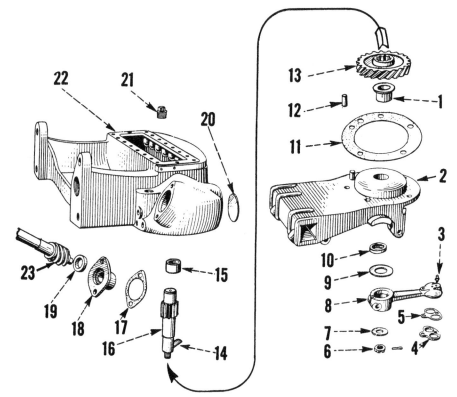

Fig. IH310—Exploded view of the Cub steering gear housing and associated parts. Housing (22) is also the radiator lower tank.

1. Bushing	8. Center steering	12. Dowel	18. Worm shaft bear-
2. Steering gear	arm	13. Worm wheel	ing
housing base	9. Thrust washer	15. Bushing	19. Oil seal
4. Steering gear	10. Oil seal	16. Worm wheel shaft	20. Expansion plug
arm plate	11. Gasket	17. Gasket	21. Oil filler plug
5. Shim			23. Worm and shaft

To remove the sector, remove hood, grille, steering gear housing cover (3) and wormshaft. Remove sector retaining nut (4) and using a suitable puller, remove sector from upper bolster pivot shaft.

Renewal of upper bolster pivot shaft oil seal (6), bushings (5) and (11) and/or thrust bearing (10), requires removal of the upper bolster pivot shaft as follows: To remove the upper bolster pivot shaft after worm shaft and sector are removed, support front of tractor and remove front axle, lower bolster or wheel fork. Withdraw the pivot shaft from below. The pre-sized bushings (5 & 11) and oil seal (6) can be renewed at this time. End play of upper bolster pivot shaft is non-adjustable; vertical thrust being taken on ball thrust bearing (10).

Series 4-6

The steering gear is of the worm and gear type mounted on rear frame (transmission case) cover. The unit is non-adjustable, except that excessive backlash between worm and worm wheel can be corrected by removing worm wheel (12—Fig. IH313) and relocating same on splines of worm wheel shaft (21).

18. **R & R AND OVERHAUL.** The gear unit can be removed from the tractor after disconnecting drag link (23—Fig. IH313), starter switch, switch box and wire harness assembly. Procedure for disassembly is self-evident after an examination of the unit and reference to Fig. IH313.

The worm shaft bushing (9) is pre-sized to an inside diameter of 1.000-1.002 inch but the two bushings (13) must be reamed or otherwise sized to 1.2505-1.2515 inches.

Series 9

The steering gear is of the cam and lever type mounted in the steering gear housing and rear frame cover. Camshaft end play and gear backlash are adjustable.

19. **ADJUST CAMSHAFT END PLAY.** This adjustment is controlled by shims (1—Fig. IH315) located between steering column (14) and steering gear housing (9). Adjustment is correct when cam (5) has zero end play and yet turns freely. To decrease end play remove shims.

20. **ADJUST BACKLASH.** The drag link must be disconnected before making this adjustment. After adjusting steering cam shaft end play, as in

paragraph 19, place gear on the high point by turning steering wheel to mid-position of its rotation. Tighten cross shaft adjusting screw (4) until a very slight drag is felt only at the mid-point when turning steering wheel slowly from the full right to the full left turn position. Gear should rotate freely at all other points.

21. **R & R AND OVERHAUL.** To remove the steering gear housing (9—Fig. IH315) and rear frame cover, first remove lighting switch, belt pulley control, magneto grounding switch, foot accelerator linkage, governor control linkage, radiator shutter control and disconnect drag link from steering gear arm (7). Remove cover to rear frame cap screws and lift the assembly from tractor.

To disassemble the unit, remove the steering arm (7) from cam lever and shaft (2) after first marking its rela-

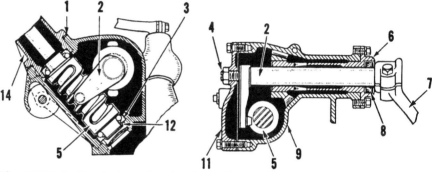

Fig. IH315—Sectional view of series 9 steering gear assembly. Cam shaft end play is controlled by shims (1).

2. Cam lever and shaft	6. Lever shaft sleeve	9. Steering gear housing
3. Bearing	7. Steering gear arm	11. Housing cover
4. Adjusting screw	8. Oil seal	12. Snap ring
5. Cam and shaft		14. Steering column

tive position on the shaft. Remove housing cover (11) and withdraw cam lever and shaft. Remove cap screws holding steering column (14) to housing (9) and withdraw steering column, cam and bearing as a unit.

When reassembling the gear unit, reinstall cam lever and shaft sleeve (bushing) (6) with the notch in flange facing up. Reinstall oil seal (8) with the lip facing inward. Use care when entering cam lever shaft into sleeve (bushing) to prevent damaging the seal.

Drag link is non-adjustable for length making it necessary to install Pitman or drop arm on cam lever shaft after gear unit is installed on tractor. With front wheels in straight ahead position and steering gear in mid-position, install steering gear arm to cam shaft lever, so as to make it unnecessary to turn either the front wheels or steering wheel to connect drag link.

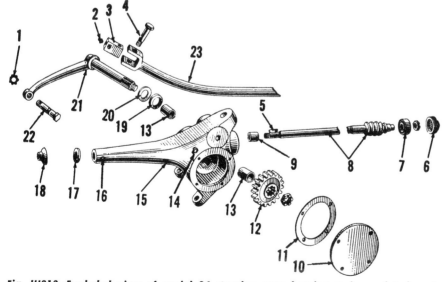

Fig. IH313—Exploded view of model 04 steering gear housing and associated parts. The gear unit as used on series 6 and other series 4 tractors is similar.

2, 14 & 16. Lubricator		18. Dust shield
3. Drag link clevis block	10. Housing cover	19. Oil seal
4. Clevis screw pin	11. Gasket	20. Felt washer
6. Bearing retainer	12. Worm wheel	21. Worm wheel shaft and arm
7. Bearing	13. Bushing	22. Gear arm screw pin
8. Worm and shaft	15. Steering gear housing	23. Drag link
9. Bushing	17. Felt washer	

ENGINE AND COMPONENTS

R&R ENGINE WITH CLUTCH

Series A-B-C-Cub

25. To remove the engine and clutch as an assembly, proceed as follows: Support tractor under clutch housing and remove radiator, steering gear unit and front axle assembly as outlined in paragraph 7, 9 or 12. On models equipped with hydraulic "Touch-Control", drain hydraulic cylinder and remove the hydraulic lines. Disconnect fuel lines, heat indicator sending unit, wiring harness and controls from engine and engine accessories. Remove oil cup from air cleaner. On all models except the Cub, remove the starting motor, remove the fuel tank front support bolts, loosen the fuel tank rear support bolts, and block-up between fuel tank and hydraulic cylinder (or clutch housing on models not equipped with "Touch-Control"). Support engine in a hoist, remove clutch housing cover, bolts retaining engine to clutch housing, and pull engine forward and away from tractor.

Series H-M

26. *There are two accepted procedures for removing the engine and clutch as a unit, from tractor: One method involves disconnecting the bolster and front end unit from side rails, removing side rails, then disconnecting engine from clutch housing. The other method involves splitting the tractor at the engine-to-clutch housing joint, then removing the engine from the front half of tractor without disturbing bolster or radiator. In either case, the preliminary work outlined in paragraph 26A is required.*

26A. Remove hood, d i s c o n n e c t steering shaft universal joint and remove the U-joint Woodruff key. Remove the radiator upper support rod and disconnect radiator hoses. Disconnect heat indicator sending unit, fuel lines, oil pressure gage line, wiring harness and controls from engine and engine accessories. Remove oil filter unit from engine and on models H and HV, remove starting motor. On Diesel models, remove the starting control rod and the Diesel fuel supply and return lines. Remove injection pump air cleaner pipe and carburetor air cleaner pipe. The remainder of the engine removal procedure differs depending on the method used. The two methods are outlined in paragraphs 26B and 26C.

26B. After performing the work which is outlined in paragraph 26A, remove dust pan from under front of frame rails, support tractor under clutch housing, remove cap screws retaining bolster and front end assembly to side rails and remove bolster. Attach hoist to engine, remove bolts retaining side rails to engine and clutch housing and remove side rails. Disconnect engine from clutch housing and remove engine.

Note: On models H, HV, Super H & Super HV, one screw which fastens clutch housing to engine end plate is accessible only through opening when starting motor is removed.

26C. After performing the work which is outlined in paragraph 26A, remove grille and steering worm shaft. Support tractor under clutch housing, and engine in a hoist. Remove bolts which retain side rails and engine to clutch housing and loosen the frame rail to bolster cap screws on one side rail.

Note: On models H, HV, Super H & Super HV, one screw which fastens clutch housing to engine end plate is accessible only through opening when starting motor is removed.

Remove engine front mounting bolts and slide the entire front unit of engine, radiator, bolster and front wheels, forward, or rear part of tractor rearward. Block front unit and slide engine rearward from the side rails.

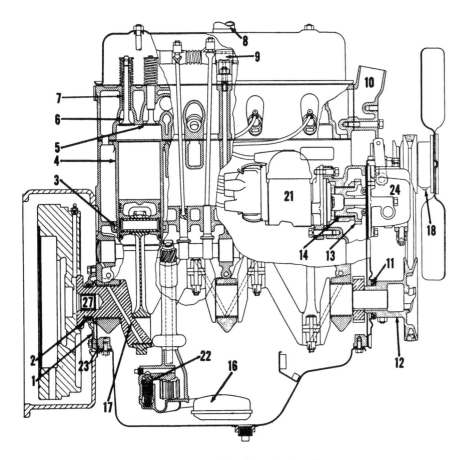

Sectional view of series A-B-C engine.

1. Oil seal retainer
2. Rear oil seal
3. Sleeve sealing ring
4. Cylinder sleeve
5. Inlet valve
6. Exhaust valve
7. Valve stem guide
8. Breather
9. Drilled lever shaft bracket
10. Combination fan bracket & water outlet
11. Crankshaft front seal
12. Crankshaft pulley
13. Magneto-governor gear
14. Drive gear bushing
16. Floto oil screen
17. Crankshaft oil passage
18. Fan oil plug
21. Magneto
22. Relief valve
23. Seal retainer plate
24. Governor housing
27. Pilot bushing

Series 4-6 (Cast Iron Frame)

27. Remove the hood, muffler, grille, and radiator. Disconnect controls and wires from engine.

On Diesel models, remove starting control rod and the Diesel fuel supply and return lines. Remove the injection pump air cleaner pipe and carburetor air cleaner pipe. Remove belt pulley unit if tractor is so equipped. Remove clutch cover and

disconnect clutch shaft coupling. Remove engine rear support cap screws and dowels (dowels can be removed by screwing a nut down on them); unbolt engine front support from front frame (be careful not to mix or lose shims which may be installed under engine supports). Support engine with hoist and lift engine from tractor frame.

Series 6 (With Channel Frame)

27A. Remove hood, disconnect radiator hoses and remove the radiator upper support rod. Disconnect heat indicator sending unit, fuel lines, oil pressure gage lines, wiring harness and controls from engine and engine

accessories. Remove oil filter unit and on Diesel models, remove the starting control rod and the Diesel fuel supply and return lines. Remove the injection pump air cleaner pipe and the carburetor air cleaner pipe.

Support tractor under clutch housing and disconnect the steering drag link and front axle stay rod. Unbolt bolster from frame rails and roll the complete front assembly away from tractor.

Attach hoist to engine in a suitable manner, remove bolts retaining side rails to engine and clutch housing and remove side rails. Disconnect engine from clutch housing and remove engine.

Fig. IH317—Rear view of Super C engine removed from tractor. The clutch cover assembly and driven plate have also been removed. Notice locking plates for the flywheel cap screws.

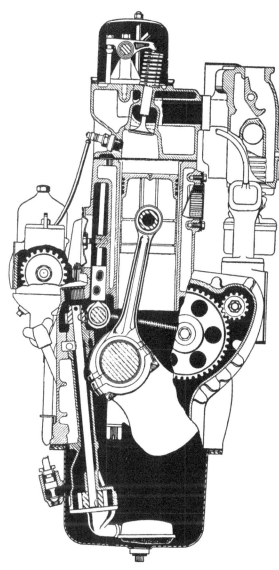

Fig. IH318—Three-quarter front view of Super MD engine. The radiator and complete front system have been removed. Late production 6 series Diesel tractors with channel type frame are similarly constructed.

Fig. IH 319—Front sectional view of series 9 gasoline engine. The engine is equipped with precision type rod bearings and three precision type mains.

Series 9

28. Remove hood, main fuel tank, battery and battery box. Remove shutter control, wiring harness and engine controls. Remove radiator brace rods and radiator shutter and housing assembly. Remove auxiliary fuel tank and air cleaner with support and gages as an assembly. On Diesel models, remove starting control rod and the Diesel fuel supply and return lines. Remove the injection pump air cleaner pipe and carburetor air cleaner pipe. Remove bolster bolts and engine front mount bolts. Remove belt pulley, clutch cover, release bearing sleeve and clutch coupling. Remove engine rear mount screws and dowels (be careful not to mix or lose shims which may be installed under engine mounts). Support engine with a hoist and lift engine from tractor frame.

CYLINDER HEAD

Series A-B-C

30. To remove the cylinder head, first drain cooling system and remove hood, valve cover, rocker arms assembly and push rods. Loosen upper radiator hose and disconnect the combination fan bracket and water outlet casting from head. Disconnect fuel lines and remove carburetor and manifold assembly. Remove the cylinder head retaining stud nuts and lift cylinder head from tractor.

When reinstalling cylinder head, tighten the stud nuts evenly and to a torque of 65 ft.-lbs.

Cub

31. To remove the cylinder head, first remove the one piece hood and fuel tank. Drain cooling system, disconnect the water outlet from front of head, remove head hold down screws and lift head from tractor.

When installing the cylinder head, tighten the cap screws evenly and to a torque of 45 ft.-lbs.

Series H-M-4-6 (Non-Diesels)

32. To remove the head, first drain cooling system and remove hood, grille, radiator brace rod and on all models except series 4 and 6, remove the steering worm shaft. Remove heat indicator sending unit, oil pressure gage line, generator and bracket and water outlet casting. Disconnect fuel lines and carburetor controls. On distillate models of the H series, remove the distillate fuel line. Remove valve cover, rocker arms assembly and push

rods. Remove cylinder head retaining stud nuts and lift cylinder head from tractor.

When installing cylinder head, tighten the stud nuts evenly and to a torque of 70 ft.-lbs. for the H and 4 series; 110 ft.-lbs. for the M and 6 series.

Series M-6 (Diesels)

33. To remove the cylinder head, first drain cooling system and remove hood, grille, radiator brace rod and on all models except series 6, remove the steering worm shaft. Remove heat indicator sending unit, oil pressure gage line and water outlet casting from cylinder head. Remove injection pump-to-nozzles lines, injection pump air cleaner pipe and carburetor air cleaner pipe. Disconnect governor arm and starting linkage and

remove the center steering shaft guide from bracket. Disconnect wiring harness and remove fuel lines and controls from carburetor. Remove valve cover, rocker arms assembly and push rods. Remove the cylinder head retaining stud nuts and using a hoist, lift cylinder head from tractor.

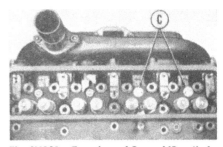

Fig. IH321—Top view of Super MD cylinder head with rocker arms assembly removed. Other M and 6 series Diesel Cylinder heads are similar.

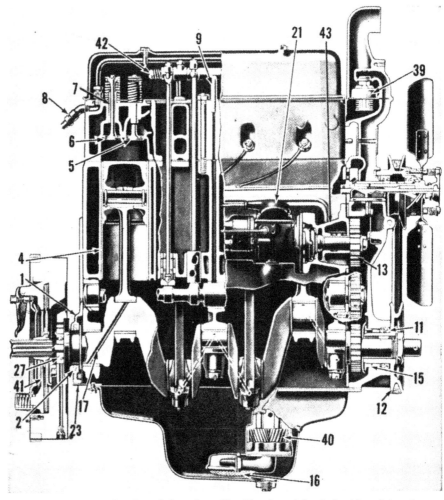

Sectional view of series H and 4 engine. Non-Diesel models of the M and 6 series are similarly constructed

1. Oil seal retainer	11. Crankshaft front seal	21. Magneto
2. Rear oil seal	12. Crankshaft pulley	23. Seal retainer plate
4. Cylinder sleeve	13. Magneto drive gear	27. Pilot bearing
5. Inlet valve	15. Oil slinger	39. Thermostat
6. Exhaust valve	16. Floto oil screen	40. Oil pump
7. Valve stem guide	17. Crankshaft oil passage	41. Clutch
8. Breather		42. Lever shaft
9. Oil sleeve		43. Magneto bracket

Before installing the cylinder head, examine both expansion plugs which are located in bottom side of head. Always renew a questionable plug. When installing head, tighten the stud nuts evenly and to a torque of 110-135 ft.-lbs.

Series 9 (Non-Diesels)

34. To remove the cylinder head, drain cooling system and remove hood, heat indicator sending unit and thermostat housing. Disconnect fuel line and carburetor controls and remove carburetor and manifold as a unit. Remove valve cover, rocker arms assembly and push rods. Remove cylinder head retaining stud nuts and lift head from tractor.

When installing cylinder head, tighten the stud nuts evenly and to a torque of 160 ft.-lbs.

Series 9 (Diesels)

35. To remove the cylinder head, drain cooling system and remove hood, heat indicator sending unit and thermostat housing. Remove injection pump-to-nozzles lines, injection pump air cleaner pipe, and carburetor air cleaner pipe. Disconnect starting linkage. Disconnect carburetor fuel line and controls and remove carburetor and manifold as a unit. Remove valve cover, rocker arms assembly and push

rods. Remove cylinder head retaining stud nuts and using a hoist, remove head.

Before installing cylinder head, examine both expansion plugs which are located in bottom side of head. Always renew a questionable plug. When installing head, tighten the stud nuts evenly and to a torque of 160 ft.-lbs.

VALVES AND SEATS

Series A-B-C-Models H-HV-O4-OS4-W4-Cub

38. Intake and exhaust valves are not interchangeable and seat directly in cylinder head (cylinder block on Cub) with a seat angle of 45 degrees and a seat width of 5/64 inch for series H, 4, A & B; 1/8 inch for series C and 3/64 inch for the Cub. On all models except the Cub, valves are equipped with safety stem retainers to prevent valve from dropping into combustion chamber. Exhaust valve seat inserts are available for service installation on the H and 4 series tractors only. Valves have a stem diameter of 0.3095-0.3105 for the Cub; 0.3405-0.3415 for other models.

Tappet gap should be set to 0.014 Hot for series A, B & C; 0.013 Cold for the Cub; and 0.017 Hot for series H and 4.

Super H-Super HV-Super W4

38A. Intake and exhaust valves are not interchangeable. Intake valves seat directly in cylinder head; whereas, the cylinder head is fitted with renewable seat inserts for the exhaust valves. Valves have a seat angle of 45 degrees and a seat width of 1/16 inch. Valves are equipped with stem retainers to prevent valve from dropping into combustion chamber. Valves have a stem diameter of 0.3405-0.3415. Tappet gap should be set Hot to 0.017 for both intake and exhaust.

Series M-6-9 (Non-Diesels)

39. Intake and exhaust valves are not interchangeable. Intake valves seat directly in cylinder head. Valves have a seat angle of 45 degrees and it is recommended that the face angle be finished to 44-44 1/2 degrees.

Intake and exhaust valve seat width is 3/32 inch for series M & 6. On series 9, intake seat width is 5/64 inch and exhaust seat width is 3/32 inch. Exhaust valve seat inserts are installed in production, and are also available for service installation. Valves are equipped with safety stem retainers to prevent valve from dropping into combustion chamber. Valves have a stem diameter of 0.371-0.372 for the M and 6 series; 0.402 for the 9 series.

Tappet gap should be set to 0.017 Hot.

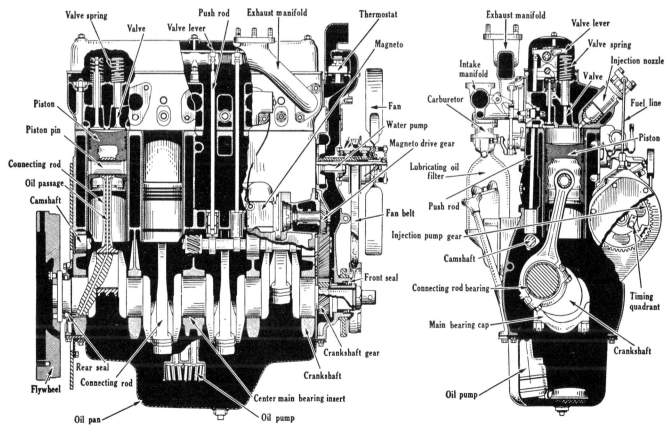

Sectional view of typical M & 6 Series Diesel engine. The 9 series Diesel engines are similarly constructed.

Series M-6-9 (Diesels)

40. **RUNNING VALVES.** Intake and exhaust valves are not interchangeable, and seat directly in cylinder head with a seat angle of 45 degrees. Intake and exhaust valve seat width is $\frac{3}{32}$ inch for series M and 6. On series 9, intake seat width is 5/64 inch and exhaust seat width is $\frac{3}{32}$ inch. Some valves are equipped with stem retainers to prevent valve from dropping into combustion chamber. Valve stem diameter is 0.371-0.372 for series M and 6; 0.402 for series 9.

Tappet gap should be set to 0.017 Hot.

41. **STARTING VALVES.** Starting valves seat directly in cylinder head with a seat angle of 45 degrees and a seat width 3/64 inch. Valves have a stem diameter of 0.309 for series M & 6; 0.312 for series 9.

Valve lobe clearance is 0.060-0.080.

VALVE GUIDES AND SPRINGS
All Models

43. Early production series A and C tractors and all series B tractors were factory equipped with renewable, shoulder type valve guides which must be pressed into the cylinder head until shoulder on guide is flush against top surface of cylinder head. Series A and C tractors beginning with the late 1951 production were factory equipped with shoulderless valve guides which must be pressed into the cylinder head until top of guide is 1 inch above valve spring seating surface on top of cylinder head. The shoulderless guides are furnished for field installation on older production series A and C tractors and all series B tractors.

On the Cub, valve guides should be pressed into the cylinder block (with necked-down portion toward tappet) until top of guide is 1 inch below gasket face of cylinder block.

On the H and 4 series, press the valve guides into the cylinder head with the chamfered end up until top of guide is 31/32 inch above valve spring seating surface on top of cylinder head. On the M, 6 and 9 series, service valve guides have a groove cut into the top outside diameter. Install the guides to the dimensions given below.

Distance from top of guide to spring seat counterbore in top of head:
Series M & 6 Non-Diesels except C264 engine.

High compression ....$1\frac{11}{64}$
Medium compression ....$\frac{25}{32}$
Low compression ....$\frac{31}{32}$
C264 engine ....$\frac{28}{32}$
Series M & 6 Diesels
Running valve guides.........$\frac{31}{32}$
Starting valve guides.........$\frac{7}{16}$

Series 9 Non-Diesels
High compression ....$1\frac{7}{32}$
Medium and low compression...$1\frac{3}{16}$
Series 9 Diesels
Running valve guides.........$1\frac{7}{32}$
Starting valve guides ....$\frac{3}{8}$
Intake and exhaust valve guides are interchangeable in any one model.

Note: Starting valve guides on Diesel models are not interchangeable with running valve guides.

On all models except the Cub, valve guides are pre-sized to provide a stem-to-guide clearance of 0.0015-0.0035 for the A, B and C series; 0.002-0.004 for the H, M, 4, 6 and 9 series. On the Cub, guides must be reamed to provide a clearance of 0.001-0.003 for the intake; 0.0015-0.0035 for the exhaust. On some models, valve guides are different lengths for high and low compression ratio engines; make certain that proper guides are installed.

44. Intake and exhaust valve springs are interchangeable in any one model except on some engines which are equipped with exhaust valve rotators.

Note: Starting valve springs on Diesel models are not interchangeable with running valve springs.

Valve springs with damper coils (closely wound coils at one end) should be installed with closed coils next to the cylinder head. Renew any spring which is rusted, discolored or does not meet the pressure test specifications given below, which apply to engines without exhaust valve rotators. For exhaust valve spring data on engines equipped with exhaust valve rotators, refer to paragraphs 51 and 53.

Valve Spring Free Length (Inches)
Series A-B-C ....$2\frac{17}{32}$
Cub ....$1\frac{31}{32}$
Super H & 4 Series..........$2\frac{9}{32}$
Other H & 4 Series...........$2\frac{17}{32}$
Super M & 6 Series
(Non-Diesels) 2 37/64
Other series M-6-9
(Non-Diesels) ....$2\frac{7}{8}$
Series M-6 (Diesel Running
Springs) ....$2\frac{11}{32}$
Series M-6-9 (Diesel Starting
Springs) ....$1\frac{31}{32}$
Series 9 (Diesel Running
Springs) ....$2\frac{7}{8}$
Test load (lbs.) @ Length (inches)
Series A-B-C ....30 @ $1\frac{59}{64}$
Cub ....23 @ $1\frac{1}{4}$
Super H & 4 Series....81-89 @ $1\frac{25}{64}$
Other H & 4 Series......30 @ $1\frac{59}{64}$
Super M & 6 Series
(Non-Diesels).......106 @ 1 35/64
Other series M-6
(Non-Diesels) ....58 @ $1\frac{25}{32}$
Series M-6 (Diesel
Running Springs) ....147 @ $1\frac{9}{16}$

Series M-6-9 (Diesel Starting
Springs) ....24 @ $1\frac{5}{32}$
Series 9 (Diesel Running
Springs) ....53 @ $2\frac{1}{2}$
Series 9 (Non-Diesels) ...53 @ $2\frac{1}{2}$

VALVE TAPPETS (CAM FOLLOWERS)

Series A-B-C-Cub

46. The mushroom type tappets operate directly in machined bores in the crankcase and can be removed after removing the camshaft as outlined in paragraph 69 or 70. Clearance of tappets in the unbushed crankcase bores should not exceed the I&T suggested limit of 0.005. Oversize tappets are not available.

Series H-M-4-6-9

47. Tappets are ported barrel type and ride directly in bores cut in the crankcase (cylinder block). Clearance of tappets in crankcase bores should not exceed the I&T suggested limit of 0.005. Oversize tappets are not available. Tappets are removed from the side of crankcase after removing valve levers (rocker arms) assembly, push rods, side cover plate (or plates) and tappet stop.

VALVE LEVERS (ROCKER ARMS)
All Models (Except Cub)

48. On series A, B and C, the valve levers and hollow shaft assembly is pressure lubricated from the center camshaft bearing via an oiler stud. The valve lever shaft diameter is 0.748-0.749.

48A. On series H and 4 and non-Diesel M, 6 and 9 tractors, the valve levers and hollow shaft assembly is pressure lubricated from the center camshaft bearing via an oil sleeve which is mounted on the valve lever shaft. When installing the valve levers and shaft assembly, the tube on the oiler sleeve must enter the reamed hole in the cylinder head. The valve lever shaft diameter is 0.748-0.749 for the H, 4, M and 6 series; 0.872-0.873 for the 9 series.

48B. On Diesel M, 6 and 9 tractors, the valve levers and hollow shaft assembly is pressure lubricated from the front camshaft bearing via an oil passage in the cylinder head. The valve lever shaft diameter is 0.872-0.873.

49. Early production tractors were equipped with cast iron valve levers which are equipped with renewable type bushings. These bushings require final sizing after installation to provide a clearance of 0.002-0.004 for the valve lever shaft. When installing the

bushings, make certain that the oil hole in bushing is in register with oil spurt hole in the valve lever. On late production tractors, the valve levers are of the welded construction with a non-serviceable bushing. The welded type lever which is available for field application on older tractors, should be renewed whenever the lever-to-shaft clearance exceeds the I&T suggested limit of 0.008.

VALVE ROTATORS
Models Super A & AV-Super C-Cub

51. Positive type exhaust valve rotator attachments ("Rotocaps") are available for both factory and field installation on gasoline burning engines of the above mentioned tractors. The attachment packages include the following:

Cub, four XCR exhaust valves, eight "Rotocap" assemblies and eight valve springs which have a free length of 1 7/16 inches. The new springs should require 14-16 lbs. to compress them to a height 1 3/16 inches.

Super A, Super AV and Super C, four "Rotocap" assemblies and four exhaust valve springs which have a free length of 2 9/64 inches. The new springs should require 30 lbs. to compress them to a height of 1 3/4 inches.

52. Normal servicing of the valve

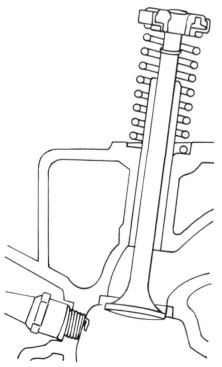

Fig. IH324—Cut away view showing Super H and 4 series exhaust valve, valve rotator and exhaust valve seat insert.

rotators consists of renewing the units. It is important, however, to observe the valve action after engine is started. The valve rotator action can be considered satisfactory if the valve rotates a slight amount each time the valve opens. A cut-a-way view of a typical "Rotocap" installation is shown in Fig. IH325.

Super H and 4 Series

52A. The above tractors are factory equipped with positive type exhaust valve rotators ("Rotocaps"). Service procedure is similar to the Super C tractor, refer to paragraphs 51 and 52. Intake and exhaust valve springs are interchangeable. Springs have a free length of $2\frac{9}{32}$ inches, and should require 81-89 lbs. to compress them to a height of 1 25/64 inches.

Series M-6-9 (Gasoline)

53. Positive type exhaust valve rotators ("Rotocaps") are factory installed on M tractors beginning with engine Ser. No. FBKM 281978, MV tractors beginning with next engine Ser. No. after FBKM 291578, W6 tractors beginning with engine Ser. No. WBKM 35937 and all Super M and 6 series tractors.

Note: The exhaust valve rotators are used only on gasoline burning engines of the above mentioned tractors.

M & 6 series springs have a free length of 2 37/64 inches and should require 102-110 lbs. to compress them to a height of 1 35/64 inches.

Series 9 springs have a free length of $2\frac{7}{8}$ inches and should require 53 lbs. to compress them to a height of 2½ inches.

Valve rotators are serviced the same as in paragraph 52.

VALVE TIMING
All Models

55. Valves are properly timed when timing marks on gears are in register. On series A, B, C, Cub, M and 6, single punch mark on camshaft gear must register with single punch mark on

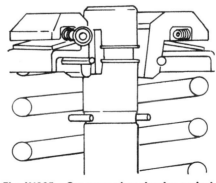

Fig. IH325—Cut away view showing typical installation of a valve rotator ("Rotocap") on series A, C, Cub, M, 6 and 9.

crankshaft gear. On series H and 4, the single marked tooth on crankshaft gear must register with the two marked teeth on camshaft gear. On series 9, single punch mark on crankshaft gear must be in register with single punch mark on idler gear and double punch marks on idler gear must be in register with double punch marks on camshaft gear.

56. To check valve timing when engine is assembled, first set the valve tappet gap to the value specified in paragraphs 38, 39 and 40, then crank engine until exhaust valve of number one cylinder is closing and intake valve of same cylinder is just beginning to open as indicated by all of the tappet gap being taken up. At this point, on series A, B and C, the "DC1-4" mark on flywheel should be approximately 1 13/32 inch or four flywheel teeth past the pointer on clutch housing cover when viewed through opening in bottom of clutch housing; on the Cub, the "DC" notch on crankshaft pulley should be approximately 13/16 inch past the pointer on the crankcase front cover; on series H, M, 4, 6 and 9, the second notch on crankshaft pulley should be in register, within ⅛ inch either way, of the pointer on the crankcase front cover.

NOTE: *The crankshaft pulley on late models of the regular H and 4 series and Super H and 4 series does not have the valve timing notch. The pulley is equipped with the notch indicating TDC only.*

TIMING GEAR COVER
Series A-B-C-Cub

58. To remove the crankcase front cover, first drain cooling system and remove complete front end assembly as outlined in paragraph 7, 9 or 12. Remove fan assembly, governor housing assembly, and on C tractors so equipped and all Super C tractors, remove water pump. Attach a suitable puller as shown in Fig. IH326 and remove crankshaft pulley.

Extra care must be taken when installing the oil seal in the crankcase front cover, so as not to distort or bend the cover. Install seal with lip of same facing inward toward timing gears.

When reassembling, leave the crankcase cover retaining cap screws loose until crankshaft pulley has been installed—this will facilitate centering the oil seal with respect to the crankshaft pulley. It is recommended that pulley be heated before driving same on crankshaft.

Fig. IH326—Puller installed for removing the crankshaft pulley on series C tractors. Series A, B and Cub are similar.

Series H-M

59. To remove the crankcase front cover, first drain cooling system and remove complete front end assembly as outlined in paragraph 15. Remove crankshaft pulley and magneto or distributor bracket and drive assembly. On Diesel models, remove generator, support front of engine and remove the engine front support. The crankcase front cover can now be removed. When reinstalling the cover, the crankshaft seal in the cover can be centered with respect to the pulley hub by turning the crankshaft several revolutions with the cover cap screws loosely installed.

Series 4-6-9

60. Remove the hood, radiator, fan blades and crankshaft pulley. Remove engine hold down bolts and block up front end of engine; then remove the engine front support (do not mix or lose mounting shims). Remove magneto or distributor bracket and drive assembly. On non-Diesels, remove carburetor and governor unit. On Diesels, remove generator and mounting bracket assembly. Cover can now be removed. When reinstalling, center crankshaft seal in same manner as described in preceding paragraph.

TIMING GEARS
Series A-B-C-Cub

62. To renew the camshaft gear and/or crankshaft gear, it is necessary to use a suitable press after the respective shafts have been removed from the engine.

The Cub idler gear which contains a bushing, rotates on a shaft which is fastened to front of crankcase by a cap screw and dowel. Clearance of the bushing on the shaft should be 0.001-0.0025.

63. When reinstalling the timing gears, mesh single punch mark on camshaft gear with single punch mark on crankshaft gear and on all models except the Cub, mesh double punch mark on camshaft gear with double punch mark on distributor or magneto-governor drive gear as shown in Fig. IH328. On the Cub, an idler gear is interposed between the crankshaft gear and the magneto or distributor-governor drive gear. See Fig. IH329. When installing the idler gear, mesh double punch mark on idler gear

Fig. IH328—Timing gear train for series B engine. Series A and C are similar.

Fig. IH329—Timing gear train for the Cub.

with double punch mark on crankshaft gear and single punch mark on idler gear with single punch mark on distributor or magneto-governor drive gear.

Note: On the Cub, the single punch marks on the distributor or magneto-governor drive gear and camshaft gear are on their rear faces.

Series H-M-4-6-9

65. The camshaft gear, crankshaft gear and/or idler gear can be removed after removing the crankcase front cover as outlined in paragraph 59 or 60. The idler gear which contains a bushing, rotates on a shaft which is fastened to front of crankcase by a cap screw and dowel. Clearance of the bushing on the shaft should be 0.001-0.0025. The idler gear on late model Diesel tractors is fitted with an anti-friction bearing.

66. When installing the timing gears on the H and 4 series tractors, mesh single marked tooth on crankshaft gear between the two marked teeth on camshaft gear; then mesh single marked tooth on camshaft gear between the two marked teeth on magneto gear; the governor gear and idler gear need not be timed to the engine.

67. When installing timing gears on non-Diesels of the M and 6 series, line up single punch mark on camshaft gear with the single punch mark on crankshaft gear; then line up double punch marks on magneto gear with the double punch marks on camshaft gear; the governor gear and idler gear need not be timed to the engine.

67A. When installing timing gears on non-Diesels of the 9 series, mesh single punch marked tooth of crankshaft gear between the punch marked teeth of idler gear; line up double punch marked tooth of idler gear with the double punch mark between teeth on camshaft gear; then line up single punch marked tooth of magneto gear with the single punch mark between teeth on camshaft gear; the governor gear need not be timed to idler gear.

68. When installing timing gears on M, 6 & 9 series Diesels, mesh single punch mark on crankshaft gear with single punch mark on idler gear; mesh double punch mark on idler gear with double punch mark on the injection pump drive gear; mesh double punch mark on idler gear with double punch mark on camshaft gear; and mesh single punch mark on camshaft gear with single punch mark on magneto or distributor drive gear.

CAMSHAFT

Series A-B-C

69. To remove the camshaft, proceed as follows: remove the complete front end assembly and the crankcase front cover as outlined in paragraph 58. Remove valve cover, valve levers and shaft assembly, push rods, oil pan and oil pump. Push tappets up into their bores. Working through openings in camshaft gear, remove the camshaft thrust plate retaining cap screws and remove camshaft from cylinder block. Camshaft gear can be removed from camshaft by using a suitable press.

Normal camshaft end play of 0.0005-0.010 is controlled by a thrust plate which is located between the camshaft gear and the crankcase. Excessive camshaft end play is corrected by renewal of the thrust plate. The camshaft rides directly in machined bores cut in crankcase. Check camshaft against the values listed below:

No. 1 (front) journal
diameter 1.811-1.812
No. 2 journal diameter. . . . 1.577-1.578
No. 3 journal diameter. . . . 1.499-1.500
Journal running clearance 0.002-0.004
Camshaft end play 0.005-0.010

Oil leakage around rear of the camshaft is prevented by an expansion plug. Renewal of this plug requires splitting the tractor as outlined in paragraph 196 and removing the flywheel.

Cub

70. To remove the camshaft, first remove engine from tractor as outlined in paragraph 25. Remove clutch, flywheel, oil pump cover over rear of camshaft and pull the oil pump drive gear from end of camshaft. Remove cylinder head, crankcase front cover and valves. Push tappets up into their bores. Working through openings in camshaft gear, remove the camshaft thrust plate retaining cap screws and remove camshaft from cylinder block. Camshaft gear can be removed from camshaft by using a suitable press.

Normal camshaft end play of 0.0003-0.012 is controlled by the thrust plate which is located between the camshaft gear and the crankcase. Excessive camshaft end play is corrected by renewal of the thrust plate. The camshaft rides directly in machined bores cut in crankcase.

Check camshaft against the values listed below:

No. 1 (front) journal
diameter 1.871-1.872
No. 2 journal diameter. . . . 1.746-1.747
No. 3 journal diameter. . . . 0.872-0.873
Journal running clearance 0.003
Camshaft end play 0.003-0.012

Series H-M-4-6-9

71. To remove the camshaft, first remove the crankcase front cover as per paragraph 59 or 60. Remove oil pan, oil pump and valve levers (rocker arms) assembly. Remove tappet cover, tappet stop and lift tappets from their bores in crankcase. Pull camshaft gear, remove the camshaft thrust plate retaining cap screws and withdraw camshaft from front of engine.

Normal camshaft end play of 0.003-0.011 for non-Diesels and 0.005-0.013 for Diesels is controlled by a thrust plate which is located between the camshaft gear and the crankcase. Excessive camshaft end play is corrected by renewal of the thrust plate. The camshaft is carried in steel-backed, babbitt-lined bushings. Renewal of the bushings requires removal of engine from tractor. If bushings are pressed or driven into position with a piloted arbor which is close fit in the bushings, they will not require reaming (check bushings for high spots after installation by installing camshaft. It may be necessary to scrape the bearings if bore is irregular). When installing bushings, the forward end of each bushing should be flush with front face of its bore and oil holes must be lined up with oil passages.

Check camshaft against the values listed below:

No. 1 (front) journal diameter
H & 4 series 1.9305-1.9315
M & 6 series
(non-Diesels). 2.243-2.244
9 series (non-Diesels). 2.3055-2.3065
M, 6 & 9 series
(Diesels) 2.4305-2.4315
No. 2 journal diameter
H & 4 series 1.8055-1.8065
M & 6 series
(non-Diesels). 2.118-2.119
9 series (non-Diesels). 2.1805-2.1815
M, 6 & 9 series
(Diesels) 2.3055-2.3065
No. 3 journal diameter
H & 4 series 1.3680-1.3690
M & 6 series
(non-Diesels). 1.868-1.869
9 series (non-Diesels). . . 1.868-1.869
M, 6 & 9 series
(Diesels) 2.1805-2.1815
No. 4 journal diameter
M, 6 & 9 series
(Diesels) 1.868-1.869
Journal running clearance 0.0015-0.0035
Camshaft end play
(non-Diesels). 0.003-0.011
Camshaft end play
(Diesels) 0.005-0.013

Oil leakage around rear of the camshaft is prevented by an expansion plug. Renewal of this plug is accomplished by first removing flywheel.

ROD AND PISTON UNITS

All Models

73. Cylinder numbers are stamped on connecting rod and cap. When installing the connecting rod and piston units, make certain that numbers on rods and caps are in register and face toward camshaft side of engine. Tighten the connecting rod bolts to the torque value specified below:
Cub. . . . . 16 ft.-lbs.
A-B-C Nuts locked with
cotter pin. . . . . 40-45 ft.-lbs.
A-B-C Self locking nuts . . 43-49 ft.-lbs.
Series H & 4 . . . . 40 ft.-lbs.
Series M & 6 (Non-Diesels). . 53 ft.-lbs.
Series M & 6 (Diesels). . . . . 115 ft.-lbs.
W9-WR9. . . . . 65 ft.-lbs.
WR9S. . . . . 87 ft.-lbs.
Series 9 (Diesels) . . . . 55 ft.-lbs.

PISTONS, SLEEVES AND RINGS

Series A-B-C

74. Iron pistons are available for standard compression ratio engines, for special compression ratio engines and engines for operation at 5000 and 8000 foot altitudes. Pistons are not available as individual parts, but only as matched units with the wet type sleeves. The matched units are available individually or in complete sets.

Reject pistons and sleeves if a spring scale pull of 11-14 lbs. will withdraw a ½ inch wide, 0.0035 thick feeler gauge.

Check new piston to new sleeve clearance, using the values listed below:
Series A-B 0.003-0.004
Feeler size . . . . ½ inch x 0.0025
Spring scale pull . . . . 4-6 lbs.
Series C 0.0022-0.003
Feeler size . . . . ½ inch x 0.002
Spring scale pull . . . . 4-6 lbs.

NOTE: *Late model pistons have the piston pin bore offset 0.065 from the piston centerline. Old style pistons with pin bore on center and new style pistons with the offset pin bore should never be used in the same engine.*

75. The wet type cylinder sleeves should be renewed if wear exceeds any of the values listed below:
Out-of-round 0.003
Taper 0.012

Special pullers are available to remove the wet type sleeves from above after the pistons have been removes. Before installing sleeves, check to make certain the counterbore at top and sealing ring groove at bottom are clean and free of foreign material. All sleeves should enter crankcase bores full depth and should be free to rotate by hand when tried in bores without sealing rings. After making trial installation without sealing rings, remove the sleeves and install new sealing rings dry into the grooves in crankcase. Wet the end of the sleeve with a thick soap solution or equivalent and install sleeves. If sealing ring is in place and not pinched, very little hand pressure is required to press the sleeve completely into place. Normally, the top of the sleeves will extend 0.003-0.007 above the machined top surface of the cylinder block. If sleeve stand out is excessive, check for foreign material under the sleeve flange.

Note: The cylinder head gasket forms the upper cylinder sleeve seal, and excessive sleeve stand out will result in coolant leakage. To test lower sealing rings for proper installation, fill crankcase (cylinder block) water jacket with cold water and check for leaks near bottom of sleeves.

76. Each piston is fitted with four rings; two plain compression rings, one tapered compression ring and one oil regulating ring. Check rings against the values listed below:

End gap (Series A-B).....0.010-0.018
End gap (Series C).......0.010-0.020
Comp. ring width
    Series A-B-Model C........⅛ inch
    Super C ....3/32 inch
Oil ring width (Kerosene) .3/16 inch
Oil ring width (Not Kerosene) ¼ inch
Top comp. ring side clearance
    Series A-B-Model C..........0.004
    Super C 0.0015-0.003

Other rings side clearance
    Series A-B-Model C..........0.003
    Super C 0.001-0.0025
Rings stamped "Top", should be installed with word top facing toward top of engine.

Compensating type replacement rings are available to correct high oil consumption without renewing the pistons and sleeves. Compensating rings are not to be used with new pistons and sleeves nor if wear exceeds any of the values listed below.

Sleeve out-of-round 0.003
Sleeve taper 0.012
Top ring clearance in groove....0.006

Cub
77. The cast iron pistons are available in standard, 0.020 and 0.040 oversize and ride directly in the unsleeved cylinder bores.

Reject pistons and/or rebore cylinder block if a spring scale pull of 11-14 lbs. will withdraw a ½ inch wide, 0.002 thick feeler gage.

Check new piston to new or newly

rebored cylinder wall clearance, using the values listed below.
Clearance 0.0016-0.0024
Feeler size ....½ inch x 0.001
Spring scale pull ....4-6 lbs.

78. Each piston is fitted with three rings; two compression rings and one oil control ring. Check rings against the values listed below.
Comp. rings end gap.....0.007-0.017
Oil ring end gap.........0.007-0.015
Comp. rings width........3/32 inch
Oil ring width............3/16 inch
Comp. rings side clearance....0.0035
Oil rings side clearance........0.003

Rings are available in standard, 0.020 and 0.040 oversize. Compensating type replacement rings are also available for use with 0.020 oversize pistons, to correct high oil consumption without renewing the cylinder block. The compensating type rings, however, are not to be used in a newly rebored cylinder block nor if piston clearances are excessive.

Series H-M-4-6-9 (Non-Diesels)
79. Pistons are available for standard compression ratio engines, for special compression ratio engines and engines for operation at 5000 and 8000 foot altitudes. Pistons are available in matched units with the sleeves. The matched units are available individually or in sets of four.

Reject pistons and sleeves if a spring scale pull of 11-15 lbs. will withdraw a ½ inch wide feeler gage of the following thickness.
Super H & 4 series 0.0025
Other series H & 4 ....0.0035
Super M & 6 series 0.003
Other M & 6 series............0.0045
W9-WR9 ....0.0065
WR9S 0.0055
Check new piston to new sleeve clearance, using the values listed below.
Super H & 4 Series......0.0021-0.0029
  Feeler size ....½ inch x 0.0015
  Spring scale pull ....4-6 lbs.
Super M & 6 Series......0.0029-0.0037
  Feeler size ....½ inch x 0.0020
  Spring scale pull..........4-6 lbs.
Other series H-4 ....0.003-0.004
  Feeler size ....½ inch x 0.0025
  Spring scale pull..........4-6 lbs.
Other Series M-6 ....0.004-0.005
  Feeler size ....½ inch x 0.0035
  Spring scale pull..........4-6 lbs.
W9-WR9 ....0.006-0.007
  Feeler size ....½ inch x 0.0055
  Spring scale pull..........4-6 lbs.
WR9S 0.005-0.006
  Feeler size ....½ inch x 0.0045
  Spring scale pull..........4-6 lbs.
80. The dry type cylinder sleeves

should be renewed if wear exceeds any of the values listed below.
Out-of-round
    Series H-M-4-6 ....0.0037
    Series 9 ....0.0044
Taper
    Series H-M-4-6 ....0.0155
    Series 9 ....0.0176

80A. To renew the dry type sleeves after pistons are out, remove them from top using IH puller SE 1213 or equivalent. Thoroughly clean crankcase bores and the counterbore at top and clean any paint or grease from the sleeves. Coat the outside of sleeves with light engine oil and press or drive sleeves into bores of crankcase. Tops of sleeves should be flush with top of crankcase but may project above it a maximum of 0.006. Top of sleeve should NOT be below top of surface of crankcase. Cylinder sleeves require no honing or boring after installation but should be checked for possible distortion or high spots.

81. Check the piston rings against the values which follow:
End gap
  Super M & 6 Series—
    Model WR9S 0.013-0.023
  Super H & 4 Series......0.007-0.017
  Other models 0.010-0.020

Top comp. ring width
  Super H-M-4-6 Series......$\frac{3}{32}$ inch
  Other H-M-4-6 Series......⅛ inch
  W9-WR9 ....9/64 inch
  WR9S ....$\frac{5}{32}$ inch

Other comp. rings width
  Super H-M-4-6 Series......$\frac{3}{32}$ inch
  Other H-M-4-6 Series......$\frac{5}{32}$ inch
  W9-WR9 ....9/64 inch
  WR9S ....$\frac{5}{32}$ inch
Oil rings width ....¼ inch
Top comp. ring side clearance...0.004
Other rings side clearance.0.002-0.0035

Rings stamped "Top", should be installed with word top facing toward top of engine. Rings having a groove in the outer face should be installed with groove down.

81A. Compensating type replacement rings are available to correct high oil consumption without renewing the pistons and sleeves. Compensating rings are not to be used with new pistons and sleeves nor if wear exceeds any of the values listed below:
Sleeve out-of-round
    Series H-M-4-6 ....0.0037
    Series 9 ....0.0044
Sleeve taper
    Series H-M-4-6 ....0.0155
    Series 9 ....0.0176
Top ring side clearance in
    groove 0.006

Series M-6-9 (Diesels)

82. Pistons are aluminum or cast iron.

Reject pistons and sleeves if a spring scale pull of 11-14 lbs. will withdraw a ½ inch wide feeler gage of the following thickness:

Super M & 6 series
　(Aluminum) 0.005
Series M-6 (Cast Iron) 0.005
Other series M-6
　(Aluminum) 0.006
Series 9 (Aluminum)........0.008
Series 9 (Cast Iron) 0.007

Check piston to sleeve clearance, using the values listed below:

Super M & 6 series
　(Aluminum) 0.0046-0.0054
　Feeler size ....½ inch x 0.004
　Spring scale pull..........3-6 lbs.
Series M-6 (Cast Iron) ..0.0045-0.0055
　Feeler size ....½ inch x 0.004
　Spring scale pull ....2-4 lbs.
Other Series M-6
　(Aluminum) 0.0055-0.0065
　Feeler size ....½ inch x 0.005
　Spring scale pull ....2-4 lbs.
Series 9 (Aluminum)....0.0065-0.0075
　Feeler size ....½ inch x 0.007
　Spring scale pull ....2-4 lbs.
Series 9 (Cast Iron).....0.0055-0.0065
　Feeler size ....½ inch x 0.006
　Spring scale pull.........2-4 lbs.
Install pistons with word "Front" toward front of engine.

83. The dry type cylinder sleeves should be renewed if wear exceeds any of the values listed below:
Sleeve out-of-round 0.0037
Sleeve taper 0.0155

Refer to paragraph 80A for renewal of the dry type cylinder sleeves. Sleeve standout above cylinder block is 0.039-0.047.

84. Check rings against the values which follow:

End gap
　Super 9 Series..........0.013-0.023
　Other Models 0.010-0.020
Top comp. ring width
　Series M-6 ....3/32 inch
　WD9-WDR9 ....⅛ inch
　Super 9 Series............$\frac{7}{32}$ inch
Second comp. ring width
　Series M-6 ....⅛ inch
　WD9-WDR9 ....$\frac{5}{32}$ inch
　Super 9 Series............$\frac{7}{32}$ inch
Third comp. ring width
　Series M-6 ....5/32 inch
　WD9-WDR9 ....$\frac{3}{16}$ inch
　Super 9 Series.........11/64 inch
Oil ring width
　Super 9 Series.....11/64 inch

Other Models ....¼ inch
Top comp. ring side clearance....0.004
Second & third comp. rings side
　clearance 0.003
Oil rings side clearance
　Series M-6 ....0.003
　WD9-WDR9 ....0.002
　Super 9 Series........0.0025-0.004

Rings stamped "Top", should be installed with word top facing toward top of engine.

Refer to paragraph 81A for precautions concerning compensating type rings.

PISTON PINS
All Models

86. The full floating type piston pins are available in standard and 0.005 oversize. Check piston pin against the values listed below.
Piston pin diameter
　Series A-B-C 0.9192-0.9195
　Cub 0.6875-0.6878
　Series H-4 ....1.1089-1.1092
　Series M-6 ....1.3125-1.3128
　Series 9 ....1.5000-1.5002
Piston pin clearance in piston
　Series A-B 0.0005
　Series C 0.0002
　Series H-M-4-6-Cub...0.0001-0.0003
　Series 9 ....0.0001-0.0005
Piston pin clearance in rod bushing
　Series A-B 0.0007
　Series C 0.0004
　Super M & 6 Series
　　Non-Diesels 0.0005-0.0007
　Series 9 ....0.0004-0.0008
　Super M & 6 Series
　　Diesels 0.003-0.005
　Other Models 0.0003-0.0005

CONN. RODS & BRGS.
All Models

88. Connecting rod bearings are of the slip-in, precision type, renewable from below. When installing new bearing shells, make certain that the rod and bearing cap numbers are in register and face toward camshaft side of engine. Bearing inserts are available in various undersizes.

Crankpin diameter
　Series A-B-C 1.749-1.750
　Cub 1.498-1.499
　H prior 391358.......2.2475-2.2485
　HV prior 391445......2.2475-2.2485
　W4 prior 34101.......2.2475-2.2485
　Other H & 4 series....2.2975-2.2985
　M prior 278050.......2.4975-2.4985
　MV prior 278107.....2.4975-2.4985
　W6 prior 35464......2.4975-2.4985
　06 to 44394..........2.4975-2.4985
　OS6 to 44461.......2.4975-2.4985
Other M-6 series
　(Non-Diesels) 2.5475-2.5485

　Series M-6 (Diesels)..3.2475-3.2485
　W9-WR9 to
　　WCBM 1020 ......2.9965-2.9975
　W9-WR9 after
　　WCBM 1020 ....2.9955-2.9965
　WR9S 2.9965-2.9975
　WD9-WDR9
　　Part No. 50977DBX..3.6230-3.6235
　WD9-WDR9
　　Part No. 260693R11..2.7480-2.7485
　Super WD9-Super
　　WDR9 ....2.7475-2.7485

NOTE: WD9 and WDR9 engines prior to WDCBM 18001 were equipped with a No. 50977BDX crankshaft which had a main journal diameter of 4.123 and a crankpin diameter of 3.623; the shaft was not counterbalanced. WD9 and WDR9 engines after WDCBM 18000 are equipped with a No. 260693-R11 crankshaft which has a main journal diameter of 4.123 and a crankpin diameter of 2.748; the shaft is counterbalanced. The two shafts are interchangeable, providing the proper size connecting rods are used.

Running clearance
　Series A-B 0.0015-0.003
　H prior 391358........0.002-0.003
　HV prior 391445........0.002-0.003
　W4 prior 34101........0.002-0.003
　Other H & 4 series.....0.0011-0.0037
　Cub 0.002-0.003
　Series C 0.001-0.0035
　M prior 278050........0.002-0.003
　MV prior 278107........0.002-0.003
　W6 prior 35464........0.002-0.003
　06 to 44394...........0.002-0.003
　OS6 to 44461.........0.002-0.003

Other M-6 series
　(Non-Diesels) 0.0011-0.0037

Super M & 6 Series
　Diesels 0.0017-0.0047

Other Series M-6
　(Diesels) 0.0023-0.0033
W9-WR9 to
　WCBM 1020 ....0.002-0.003
W9-WR9 after
　WCBM 1020 ....0.002-0.0045
WR9S 0.002-0.005
WD9-WDR9 ....0.0025-0.0035
Super WD9-
　Super WDR9 ....0.0019-0.0049

Side Play
　Series A-B-C 0.003-0.013
　Cub 0.005-0.012
　H prior 391358.........0.008-0.012
　HV prior 391445.........0.008-0.012
　W4 prior 34101.........0.008-0.012
　Other H & 4 series......0.005-0.012
　M prior 278050.........0.008-0.012
　MV prior 278107.........0.008-0.012
　W6 prior 35464........0.008-0.012

06 to 44394..............0.008-0.012
OS6 to 44461..............0.008-0.012
OS6 - 44461..............0.008-0.012
Other M-6 series
  (Non-Diesels) 0.005-0.012
Series M-6 (Diesels) 0.003-0.010
W9-WR9 ....0.008-0.015
WD9-WDR9 ....0.003-0.010
WR9S 0.009-0.015
Super WD9-Super
  WDR9 ....0.009-0.015

Rod bolt torque (ft.-lbs.)
A-B-C Nuts locked with
  cotter pin 40-45
A-B-C Self locking nuts.....43-49
Series H-4 40
Cub 16
Series M-6 (Non-Diesels).......53
Series M-6 (Diesels)...........115
W9-WR9 65
WR9S 87
Series 9 (Diesels) 55

CRANKSHAFT AND BEARINGS
All Models

90. The crankshaft is supported in three main bearings on non-Diesel engines; five main bearings on Diesel engines. End thrust is taken on the rear main bearing on series A, B and C; center main bearing on all other models. Main bearings are of the shimless, non-adjustable, slip-in, precision type, renewable from below after removing the oil pan and main bearing caps. Removal of the rear main bearing cap on all models except the Cub, requires removal of the crankshaft rear oil seal lower retainer plate. Renewal of crankshaft requires R & R of engine. Check crankshaft and main bearings against the values listed below.

Crankpin diameter........
       ....Refer to paragraph 88

Main journal diameter
Series A-B-C 2.124-2.125
Cub 1.623-1.624
H prior 391358.......2.4975-2.4985
HV prior 391445......2.4975-2.4985
W4 prior 34101.......2.4975-2.4985
Other H & 4 series.....2.5575-2.5585
M prior 278050.......2.7475-2.7485
MV prior 278107......2.7475-2.7485
W6 prior 35464.......2.7475-2.7485
O6 to 44394.........2.7475-2.7485
OS6 to 44461.........2.7475-2.7485
Other M-6 series
  (Non-Diesels) 2.8075-2.8085
Series 9 (Non-Diesels).3.2475-3.2485
Series 9 (Diesels).....4.1225-4.1235
Series M-6 (Diesels)..3.7475-3.7485

Crankshaft End Play
WR9S 0.006-0.010
WD9-WDR9 ....0.002-0.010
Other Models 0.004-0.008

Main bearing running clearance
Series A-B 0.0015-0.003

Series C 0.001-0.0035
Cub 0.002-0.0035
H prior 391358.........0.002-0.003
HV prior 391445........0.002-0.003
W4 prior 34101.........0.002-0.003
Other H & 4 series.....0.0011-0.0037
M prior 278050.........0.002-0.003
MV prior 278107........0.002-0.003
W6 prior 35464.........0.002-0.003
06 to 44394............0.002-0.003
OS6 to 44461...........0.002-0.003
Other M-6 series
  (Non-Diesels) 0.0011-0.0037
Super M & 6 Series
  Diesels 0.0018-0.0048
W9-WR9 ....0.002-0.003
WR9S 0.0019-0.0049
Super WD9-Super
  WDR9 ....0.002-0.005
Other Series M-6-9
  (Diesels) 0.0027-0.0037

Main bearing bolt torque (ft.-lbs.)
Series A-B-C-H-4 75
Cub 55
Series M-6 (Non-Diesels)......100
Super WD9-Super
  WDR9 (center)..............250
Super WD9-Super
  WDR9 (others)..............150
Other Series M-6-9
  (Diesels) ⅝ studs........150-175

Other Series M-6-9
  (Diesels) ¾ studs........250-275
Series 9 (Non-Diesels) 125

Main bearings are available in standard as well as various undersizes.

CRANKSHAFT REAR OIL SEAL
All Models

The rear oil seal on late production Cub tractors and for replacement purposes on earlier models, is rubber coated to reduce the possibility of seal becoming loose in the seal retainer.

92. **RENEW.** Procedure for renewal of the crankshaft rear oil seal is evident after removing the flywheel as outlined in paragraph 94.

When installing rubber coated seals on Cub tractors, coat the rubber covering with chassis lubricant to prevent damage. Note: If old style (not rubber covered) seal is being replaced by the new (rubber covered) seal, it is also necessary to install the new style seal retainer (IH part No. 251363 R12).

On M, 6 & 9 series tractors when installing the late style rubber seal package, be sure to follow instructions in seal package for reworking crankshaft.

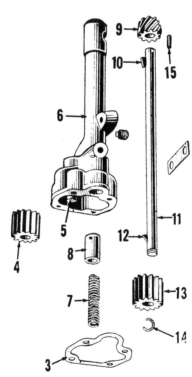

Fig. IH332—Exploded view of series A-B-C oil pump. Series H-4 and M-6 (Non-Diesels) are similar except the drive gear is retained by a pin.

1. Screen
3. Gasket
4. Idler gear
5. Idler gear shaft
6. Pump body
7. Relief valve spring
8. Relief valve
9. Drive pinion
10. Key
11. Drive shaft
12. Key
13. Body drive gear
14. Retainer
15. Pin

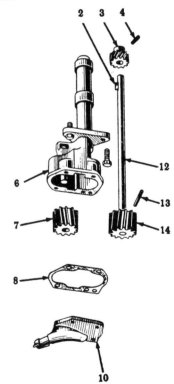

Fig. IH333—Exploded view of series 9 Diesel engine oil pump. Series 9 Non-Diesel and series M & 6 Diesels are similar. The oil pressure relief valve is located in the oil filter base.

2. Key
3. Drive pinion
4. Pin
6. Pump body
7. Idler gear
8. Gasket
9. Screen
10. Cover
12. Drive shaft
13. Pin
14. Body drive gear

FLYWHEEL

All Models

94. The flywheel can be removed on series A, B, C, H, M, Cub and 4 & 6 with channel frame after splitting tractor at clutch housing. On series 9, the clutch compartment cover and clutch must be removed. On series 4 and 6 with cast iron frame, the engine must be removed from tractor.

To install the flywheel ring gear, heat same to approximately 500 deg. F. and install gear on flywheel so that beveled end of the ring gear teeth will face beveled end of teeth on starting motor pinion.

OIL PUMP

Series A-B-C-H-M-4-6-9

96. The gear type oil pump, which is gear driven from a camshaft pinion, is accessible after removing the engin oil pan. Disassembly and overhaul of the pump is evident after an examination of the unit and reference to Fig. IH 332 or 333. Gaskets between pump cover and body can be varied to obtain the recommended gear end play. Check the pump parts against the values listed below.

Diametral clearance of gears in pump
body
  Series A-B-C-H-4 ......0.005-0.008
  Series M-6 (Non-
    Diesels) 0.004-0.006
  WR9S 0.012-0.015
  Super WD9-Super
    WDR9 ....0.003-0.0075
  Other Series 9...............0.003
  Series M-6 (Diesels) ..0.006-0.0075
Body gears end play
  Series A-B-C 0.0035-0.006
  Series H-4-M-6 ....0.003-0.006
  Series 9 (Non-Diesels)...0.003-0.006
  WD9-WDR9 ....0.005-0.010

Fig. IH335—Cub oil pump is accessible after splitting engine from clutch housing and removing the flywheel.

A. Pump idler gear
B. Idler gear stud
C. Pump drive gear on end of camshaft
D. Crankshaft rear oil seal retainer
E. Oil pump body
F. Pump gasket

Super WD9-Super
  WDR9 ....0.003-0.006
Body gears backlash
  Series 9 ....0.004-0.006
  Other Models 0.003-0.006

Cub

97. As shown in Fig. IH335, the gear type oil pump which is bolted to the rear face of the engine is accessible after splitting the tractor at the clutch housing and removing the flywheel.

The drive gear (C) is pressed and keyed to the camshaft. Idler gear (A) rotates on a stud (B) which is pressed into a bore in the crankcase. The oil pump body (E) is retained to the crankcase by cap screws. To renew the idler gear stud (B) first remove camshaft as outlined in paragraph 70 and, using a long punch inserted through front and center camshaft bearing holes, drive stud from crankcase. Gaskets (F) between crankcase and pump body can be varied to obtain end play of gears. Check pump parts against the values listed below.

Diametral clearance of gears in
  body 0.008-0.012
Body gears end play 0.002-0.006
Body gears backlash 0.007-0.012
Oil pump pickup tube and intake

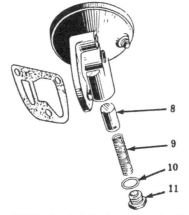

Fig. IH336—Exploded view of series 9 oil filter base, showing the relief valve location.

8. Relief valve
9. Spring
10. Gasket
11. Nut

screen is accessible after removing oil pan.

OIL PRESSURE RELIEF VALVE

All Models

99. On all models, the spring loaded plunger type oil pressure relief valve is non-adjustable. On the Cub, the valve is located in the side of the crankcase, just below the base for the magneto. On all series 9 tractors and Diesel models of the M and 6 series, the valve is located in the oil filter base as shown in Fig. IH336. On all other models, the valve is contained in the oil pump body as shown in Fig. IH332. Check oil pressure and relief valve against the values listed below.

Oil pressure-psi
  Series A-B-C 50-60
  Cub 30-35
  Super M-6-9 Series Diesels....38-46
  Other Models 60-70

Relief valve diameter
  Series H-M-4-6-9 ......0.900-0.901

Relief valve clearance in bore
  Series H-M-4-6-9 ......0.004-0.006

Relief valve spring test data
  Series A-B-C ..26 lbs. @ $1\frac{5}{8}$ inches
  Cub ......$9\frac{1}{2}$ lbs. @ 2 15/32 inches
  Series 9 Non
    Diesels..38.2 lbs. @ 2 3/32 inches
  Super M & 6
    Diesels .....27 lbs. @ $2\frac{3}{32}$ inches
  Super WD9-Super
    WDR9......19.5 lbs. @ $2\frac{3}{32}$ inches
  Other Models..42 lbs. @ $2\frac{3}{32}$ inches

CARBURETOR

Series A-B-C

101. Series A, B and C tractors are equipped with either Carter, Marvel-Schebler or Zenith carburetors as listed in an accompanying table.

Table 1
Carburetor Applications

	Carter	Marvel Schebler	Zenith
A-AV-B-BN Gas	—	TSX-157	161X7-9752
A-AV-B-BN Ker. or Dist.	—	TSX-156	161X7-9749
Super A&AV Gas	UT733S or SA	TSX-157	161X7-10514
Super A&AV Ker. or Dist. (early)	UT734S	TSX-156	161X7-10522
Super A&AV Ker. or Dist. (late)	—	—	67X7
C Gas	UT733S or SA	TSX-319	161X7-10386
C Ker. or Dist.	UT734S	TSX-333	161X7-10441
Super C Gas	UT771SA, 733S or SA	—	67X7
Super C Ker. or Dist.	UT925S	—	67X7

Series H-4-6-9-Cub-M (except L.P. Gas Models)

102. The above models are equipped with International Harvester carburetors as follows:

Cub ....¾ Updraft
Super H & 4 series.............E 12
Other series H-4 ....D 10
Series M-6 (Non-Diesels).......E 12
Series 9 (Non-Diesels).........E 13
Series M-6-9 (Diesel Starting)..F 8

103. **MODELS ¾ UPDRAFT, D10, E12, & E13.** These carburetors are of the plain tube updraft type and, with the exception of the ¾ updraft type used on the Cub, are made in two cali-

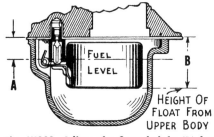

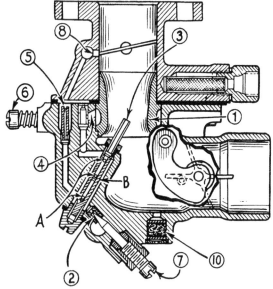

Fig. IH339—Adjust the float height (B) for IH carburetors to obtain fuel level (A).

Fig. IH340—Sectional view of IH D10 carburetor.

brations; one for use with gasoline burning engines, and one for use with kerosene or gasoline-distillate burning engines.

104. ADJUSTMENTS. All models are equipped with an air controlling idle adjusting screw which reduces the amount of air and richens the mixture when turned clockwise. All models except the ¾ inch updraft which is used on the Cub, are equipped with fuel controlling main jet adjusting screw which leans the mixture when turned clockwise. Adjust the float height (Fig. IH339) to the dimension which follows.

¾ Updraft ....$1\frac{13}{32}$
D10 ....$1\frac{53}{64}$
1¼ Updraft ....$1\frac{5}{16}$
E12-E13 ....$1\frac{5}{16}$

105. OVERHAUL NOTES. When overhauling the carburetor (Fig. IH 340 or 341), disassemble it completely, clean and inspect all parts for wear. Some of the units have renewable throttle shaft bushings. On unbushed units, excessive wear is corrected by renewing the throttle shaft and/or throttle body. Inspect the idle adjusting screw, fuel adjusting screw and seat and inlet needle and cage for ridges or depressions. Check for excessive wear in the float pivot pin and support and for burrs on choke and throttle valve. Inspect drip hole filler and renew the filler if it is not in good condition. Refer to IH parts catalogs for calibration data.

When assembling model E12 or E13 carburetor, the bushing nearest the throttle stop must be installed as shown in Fig. IH342 as air in the economizer slot passes through the holes in the bushing. Install the throt-

A. Holes in discharge nozzle
B. Metering well
1. Venturi
2. Main jet
3. Discharge nozzle
4. Main air bleed
5. Idling jet
6. Idle air adjusting needle
7. Main jet adjusting screw
8. Idling slot
10. Drip hole filler
11. Economizer slot (E12-E13 only)

tle butterfly so that the side marked 12° faces up, when throttle is closed. With throttle fully closed (stop screw backed out), 0.025-0.031 of the idle slot should be exposed above the throttle butterfly as shown in Fig. IH343. If this condition is not obtained, check for wear in the barrel, butterfly, throttle shaft and/or bushings.

106. **MODEL F8.** A cross-sectional view of the F8 carburetor, which is used for starting the Diesels engines only, is shown in Fig. IH345. When

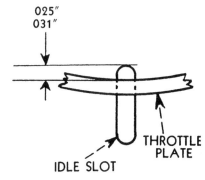

Fig. IH343—Setting throttle butterfly plate in relation to idle slot in models E12 & E13 carburetor.

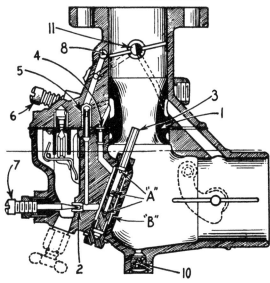

Fig. IH341—Sectional view of IH E12 & E13 carburetors.

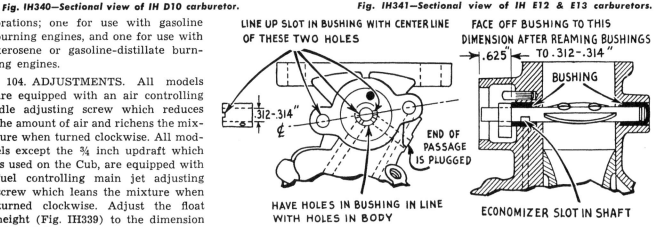

Fig. IH342—Installing throttle shaft bushings in IH Models E12 & E13 carburetors.

the engine is starting (running on gasoline), the cam (2) is in the position shown. When switching from gasoline to Diesel operation, the cam contacts spring (3); therefore, holding the inlet needle valve (4) against its seat. High velocity air entering around choke valve (7) passes around air valve (8) which is open at all times. The vacuum around air valve (8) draws fuel from nozzle (13).

107. ADJUSTMENTS. Early production model F8 carburetors are provided with an air valve adjustment and a fuel level (or float) setting. Late production carburetors are provided with a fuel level (or float) setting only.

To adjust the air valve on early models, remove cotter pin (19—Fig. IH345) and turn collar (20) to obtain the proper distance between air valve (8) and upper body (15). The proper distance, which must be the same all the way around the air valve, is 3/32-7/64 inch for M and 6 series tractors; 7/64-⅛ inch for 9 series tractors.

Obtain the recommended fuel level of 13/32-7/16 inch (float setting of 9/32 inch) below the bowl rim by bending the float tang or by varying the number of gaskets under the inlet needle cage.

108. OVERHAUL NOTES. Disassembly and inspection of the carburetor is evident after an examination of the unit and reference to Fig. IH345. Install air valve guide so that top of guide is 1¼ inches above carburetor body. The air valve spring has a free length of 9/16 inch and should require 3 lbs. to compress it to a height of ⅜ inch.

Note: Late production F8 carburetors are not equipped with the air valve guide and spring.

L. P. GAS SYSTEM

Super M & 6 Series

The Super M & 6 series tractors are equipped with an Ensign model XG updraft type gas carburetor which has a diaphragm type economizer. The carburetor has three adjustments; main (load) adjustment, throttle stop screw and starting adjustment. The system also includes a removable cartridge type fuel filter which can be disassembled and cleaned, and an Ensign model W regulating unit which is fitted with an idle adjusting screw.

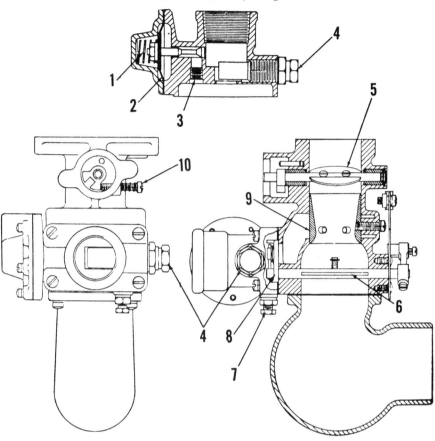

Fig. IH348—Sectional views of Ensign model XG carburetor for LP gas which is used on the Super M and 6 series.

1. Economizer spring	4. Load adjustment	7. Starting adjustment	9. Venturi
2. Economizer diaphragm	5. Throttle valve	8. Valve	10. Throttle stop screw
3. Orifice	6. Choke valve		

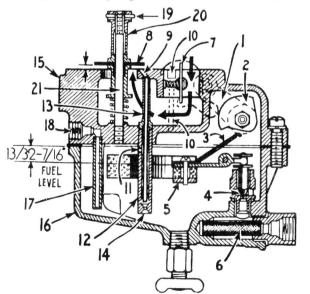

Fig. IH345 — Sectional view of early production IH model F 8 carburetor which is used for starting the Diesel engines. Late production carburetors are similar.

1. Locking lever
2. Cam
3. Spring
4. Needle valve
5. Float
6. Screen
7. Choke valve
8. Air valve
9. Fuel passage
10. Air passage
11. Air bleed
12. Metering well
13. Nozzle
14. Jet
15. Upper body
16. Lower body
17. Fuel supply tube
18. Fuel inlet
19. Cotter pin
20. Air valve collar
21. Guide

110. **ADJUSTMENTS.** Before attempting to adjust the system, start the engine and bring to operating temperature. The load adjustment should be made when engine is running at approximately 1600 rpm. Clockwise rotation of the load adjustment screw (4—Fig. IH348), which controls the amount of fuel, will lean the mixture. Retard the engine speed control and adjust the throttle stop screw (10) to obtain an engine speed of approximately 450 rpm; then, turn the idle adjusting screw (1—Fig. IH 349) to obtain a smooth idle. Clockwise rotation of the idle adjusting screw will enrich the mixture.

Set the starting adjustment (7—Fig. IH348) two-thirds turn open, which should be satisfactory for average fuel and operating conditions. If engine is difficult to start, set the starting adjustment with engine cold as follows: Close choke, start engine, open the throttle ½ way and turn screw (7) to obtain maximum engine speed.

Note: The above mentioned adjustments should provide satisfactory results for average fuel and operating conditions. If satisfactory conditions are not obtained, refer to Massey-Harris Shop Manual (Model 44LP) for complete adjustment procedure.

111. CARBURETOR OVERHAUL. The carburetor is serviced the same as a conventional gasoline type carburetor; that is, the carburetor can be completely disassembled, cleaned and worn parts can be renewed. Make certain, however, that vacuum connections to the economizer chamber do not leak.

112. REGULATOR TROUBLE-SHOOTING. If engine will not idle properly, and if turning the idle adjusting screw will not correct the condition, it will be necessary to disassemble the regulator unit and thoroughly clean the low pressure valve (7—Fig. IH349).

To test the condition of the high pressure valve (5), install a suitable pressure gage at the pipe plug con-

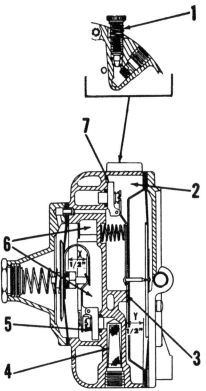

nection in the face of the regulator body. If the gage pressure increases after a warm engine is stopped, the high pressure valve is leaking. Under normal operating conditions, the high pressure valve should maintain a pressure of 3½-5 pounds. If the valve leaks, or does not maintain the proper pressure, disassemble the unit and renew the high pressure valve.

If, after standing for some time, the regulator unit is cold and shows moisture and frost, either the high or low pressure valve is leaking or the valve levers are not properly set.

113. REGULATOR OVERHAUL. Disassembly of the regulator is evident after an examination of the unit and reference to Fig. IH349. Thoroughly clean all parts and renew any which are excessively worn. When reassembling the unit, make certain that the valve levers are set to the dimensions (x & y), as shown.

Fig. IH349—Sectional views of Ensign model W regulating unit as used on LP gas burning models of the Super M and 6 series.

1. Idle adjusting screw	5. High pressure valve
2. Low pressure chamber	6. Vaporizing chamber
3. Boss	7. Low pressure valve
4. Strainer	

DIESEL SYSTEM

When servicing any unit associated with the fuel system, the maintenance of absolute cleanliness is of utmost importance.

Several different International Harvester and American Bosch injection pumps have been used. Field change-over packages are available from International Harvester Co. to change from a Bosch system to an IH type A system, and to change from the old style IH system to the type A. Due to the availability and general overall usage of the field change-over packages, only the late model, International Harvester injection system will be treated in this section. Some very late models are equipped with a type B injection pump. Type B pumps are similar to the type A except the scavenging pump is not used.

DIESEL SYSTEM TROUBLE SHOOTING CHART

	Lack Of Fuel	Engine Surging Or Rough	Cylinders Uneven	Engine Smokes Or Knocks	Inj. Pump Does Not Shut Off	Engine Dies At Low Speed	Inj. Pump Sump Dilution	Loss Of Power
Defective speed control linkage......	★				★			
Air in fuel system..................	★							
Clogged filters and/or water trap....	★							★
Fuel lines leaking or clogged........	★	★	★				★	★
Friction between plunger and rack...		★						★
Inferior fuel.......................		★						★
Faulty injection pump timing........		★		★		★		★
Loose nut on camshaft..............		★						
Defective nozzle		★	★	★				★
Faulty plunger tappet..............								★
Faulty governor and/or linkage adjustment		★		★	★			★
Scored or dirty plunger.............						★		★
Defective camshaft oil seal..........							★	
Faulty scavenging valve.............							★	
Faulty primary pump and/or pump gaskets							★	★
Faulty scavenging pump.............							★	
Clogged breather							★	
Dirty fuel return check valve.......							★	
Faulty distributor block............	★	★	★	★				

121. **QUICK CHECK-UNITS ON TRACTOR.** If the Diesel engine does not start or does not run properly, and the Diesel fuel system is suspected as the source of trouble, refer to the Diesel System Trouble Shooting Chart and locate points which require further checking. Many of the chart items are self-explanatory; however, if the difficulty points to the fuel filters, injection nozzles and/or injection pump, refer to the appropriate sections which follow:

FILTERS AND BLEEDING
Series M-6-9

122. As shown in Figs. IH352, 353 & 354, the fuel filtering system consists of a water trap and two stages of renewable element type filters.

123. **MAINTENANCE.** Start engine and note position of hand on the fuel pressure gage. If hand is in the "RED" area, remove the water trap and thoroughly clean same. Reinstall the water trap, bleed the system as outlined in paragraph 124 and recheck the pressure gage reading. If the gage hand is still in the "RED", renew the auxiliary fuel filtering element and clean the primary pump filtering screen. To renew the element, remove the filter case and element and thoroughly clean the case and filter base. Install plates (4 & 6—Fig. IH353) on new element, making certain that the plates slide into wire coil inside of the element and that face of top plate

is installed with word "TOP" toward top of tractor. Remove the primary pump filter screen as shown in Fig. IH356 and thoroughly clean same using kerosene or clean Diesel fuel. Reassemble the parts, bleed the system and recheck the pressure gage reading. If gage hand is still in the "RED", it will be necessary to renew the final fuel filtering element. The procedure for renewing the final filtering element is evident after an examination of the unit and reference to Fig. IH 353.

CAUTION: The filter case studs (2) are not interchangeable and must be installed in their original position.

If after renewing the final fuel filtering element, the pressure gage hand is still in the "RED", it will be necessary to renew or recondition the primary pump as outlined in paragraph 135 or 136.

124. **BLEEDING.** The fuel system should be purged of air whenever the fuel filters have been removed or when the fuel lines have been disconnected. To bleed the system, first make certain that fuel level in tank is above the auxiliary fuel filter, open the fuel tank shut-off valve and proceed as follows: Open the water trap vent (1— Fig. IH352) and the auxiliary fuel filter vent cock (2); close the vents when the fuel flows free of air. Start the engine and run on the gasoline cycle. Open the final fuel filter vent

cock (3); close the vent when fuel flows free of air. Advance the engine speed control lever slightly but do not touch the compression release lever. Open each nozzle vent (4) individually; close the vent when the fuel flows free of air.

INJECTION NOZZLES
Series M-6-9

WARNING: Fuel leaves the injection nozzles with sufficient force to penetrate the skin. When testing, keep your person clear of the nozzle spray.

126. **TESTING** If the engine does not run properly, or not at all, and the quick checks outlined in para-

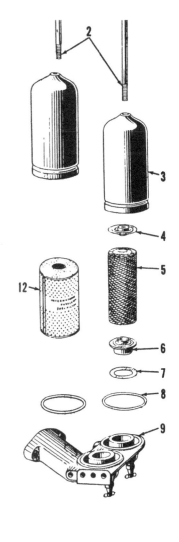

Fig. IH353—Exploded view of IH fuel filters. Case studs (2) are not interchangeable.
1. Bleeder valve
5. Auxiliary filter element
6. Filter element bottom plate
7. Gasket
8. Gasket
9. Filter base
12. Final filter element

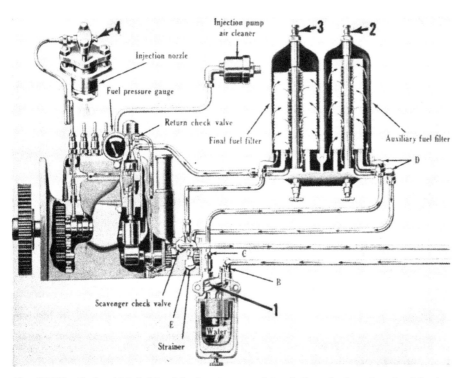

Fig. IH352—Series M-6-9 Diesel fuel system consists of three basic units; the injection nozzles, injection pump and fuel filters.

1. Water trap vent   2. Auxiliary filter vent   3. Final filter vent   4. Nozzle vent

graph 121 point to a faulty injection nozzle, test the nozzles as follows: Start engine and run on Diesel cycle. Open and close the nozzle bleeder valves (4—Fig. IH352) one at a time until the cylinder (or cylinders) is found that least affect engine performance. If a cylinder is misfiring, it is reasonable to suspect a faulty nozzle; however, a similar condition can be caused by a defective distributor valve or spring (Refer to paragraph 142).

127. R & R AND OVERHAUL. To remove the nozzle, first remove dirt from nozzle and cylinder head and disconnect the injector pipe. Remove the two stud nuts which retain nozzle to cylinder head and lift nozzle unit from head. Refer to Fig. IH358.

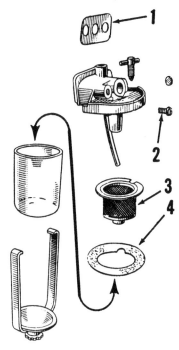

Fig. IH354 — Exploded view of IH water trap. It is necessary to bleed the fuel system after cleaning the water trap.
1. Gasket
2. Vent screw
3. Screen
4. Gasket

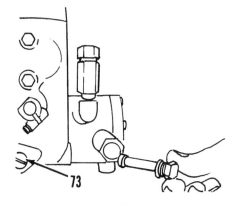

Fig. IH356—Removing the IH primary pump filter screen. Zero or low fuel pressure can be caused by a clogged screen.

NOTE: Unless complete nozzle testing equipment is available, there is little actual overhaul work which can be accomplished on the nozzle; however, the nozzle can be partially disassembled and cleaned, and some of the parts can be renewed without the use of special testing equipment. Disassemble and clean the nozzle as follows:

127A. Remove cap screws retaining the nozzle fitting to the nozzle body and remove the fitting. Remove the valve spacer and gaskets. Remove the nozzle valve and spring unit, but do not attempt to disassemble the unit. Remove injector plate and gasket. Thoroughly clean and inspect all parts, and renew any which are damaged. The nozzle opening pressure is controlled by the adjustment of the nozzle valve and spring. This adjustment should not be disturbed unless nozzle testing equipment is available. If such equipment is not available and the valve assembly is suspected as the source of trouble, install a new nozzle valve which is available as a pre-adjusted unit. If nozzle testing equipment is available, refer to paragraph 127B for overhaul of the nozzle valve.

Inspect the orifice size in the injector plate. The orifice diameter should be the value which is etched on the side of the plate. Always renew a questionable orifice plate. NOTE: The orifice diameter can be checked with the piano wire which has the same diameter as that specified for the orifice.

Assemble the nozzle in the sequence shown in Fig. IH358 and tighten the nozzle fitting cap screws to a torque of 30-35 ft.-lbs. Install nozzle by reversing the removal procedure and tighten the nozzle retaining stud nuts to a torque of 20-25 ft.-lbs. on the M and 6 series; 45-50 ft.-lbs. on the 9 series.

127B. OVERHAUL NOZZLE VALVE. As previously mentioned, the nozzle valve can be overhauled and the nozzle opening pressure can be adjusted, providing complete nozzle testing equipment is available.

Disassemble the nozzle valve and thoroughly clean parts in Diesel fuel. Inspect face and seat of nozzle valve with a magnifying glass. If face and seat are not excessively worn, they can be reconditioned by lapping by using carborundum H-400 fine until polished surfaces are obtained. Check

the nozzle valve spring against the values listed below.
Free length 0.802-0.834
Test length ....⅝ inch
Test load (lbs.) 33.06-36.54

Renew the nozzle valve spring if it is rusted, discolored or if it does not meet the foregoing pressure specifications.

Reassemble the nozzle valve and nozzle and place nozzle in a test fix-

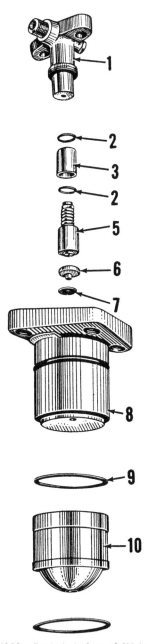

Fig. IH358—Exploded view of IH injection nozzle. The nozzle valve (5) can be overhauled providing testing equipment is available. Some late models are equipped with a nozzle fitting dust seal.
1. Nozzle fitting
2. Valve gasket
3. Spacer
5. Injector plate
7. Gasket
8. Nozzle body
9. Precombustion chamber gasket
10. Precombustion chamber

ture. Check nozzle against the values listed below.

Opening pressure (new
   spring) .... 740-750 psi
Opening pressure (used
   spring) .... 690-710 psi
No leakage below .... 400 psi

If opening pressure is too low, disassemble the nozzle and tighten the nozzle valve adjusting nut; if pressure is too high, loosen the nut. If leakage occurs below 400 psi, the nozzle valve and seat requires renewal or further lapping.

PRECOMBUSTION CHAMBERS
Series M-6-9

129. **REMOVE AND REINSTALL.** The precombustion chambers can be pulled from cylinder head after removing the respective nozzle assembly. Pre-cup Puller No. 1 020 233 R91 will facilitate removal of the precombustion chamber.

When installing the precombustion chamber, make certain that side stamped "TOP" is installed toward top of engine.

INJECTION PUMPS
(IH BUILT PUMPS)
Series M-6-9

The Diesel injection pump is located on the left side of the cylinder block. The pump, which is flange mounted to the crankcase front cover plate, is gear driven from the timing gear train.

The engine speed governor and weights unit is located below the pump camshaft in the lower half of the pump housing. The governor weights are on the front of the governor shaft and the eccentric which operates the single plunger is located on the rear of the governor shaft.

130. **TIMING.** For normal operation, injection should occur 12½ degrees before top center. It may, however, be necessary to change this timing slightly to obtain the desired running conditions for various load requirements and different grades of fuel. For any given load condition and grade of fuel, the optimum timing is when engine speed is maximum with a clean exhaust.

Note: The following timing procedure is outlined on the assumption that the injection pump drive gear is properly timed with respect to the timing gear train idler gear. Refer to paragraph 68.

130A. To check and adjust the injection pump timing when pump is installed, first crank engine until No. 1 cylinder is coming up on its compression stroke and notch "DC" on crankshaft pulley is in register with the pointer on the crankcase front cover. Remove the injection pump drive gear cover from front of timing gear case. The pump is properly timed to the pump drive gear if the chisel marked gear hub groove (A—Fig. IH359) is in line (or within 20 degrees either way) with the zero degree graduation on the injection pump drive gear as shown. If the timing marks are not as specified, it will be necessary to re-position the drive gear hub, with respect to the drive gear, as follows: Remove the three cap screws (B) and turn the drive gear hub (C) until chisel marked groove in the hub is in the position outlined above. Position the timing indicator (D) so that pointer is in register with the zero degree graduation on the gear and install the cap screws (B).

Start the engine and check the engine speed. Stop the engine and shift the timing pointer either way and recheck the engine speed. Continue this operation until maximum engine speed is obtained for any given load, and engine operation is smooth with a clean exhaust.

132. **TESTING.** If the engine does not run properly, or not at all, and the quick checks outlined in paragraph 121 point to a faulty injection pump, test the pump as outlined in paragraph 132A if test stand is available or 132B if test stand is not available.

132A. Remove the complete injection pump from tractor as outlined in paragraph 133, place pump on a test stand, operate pump at 550 rpm and check the pump against the specifications which follow:

Note: Leakage tests should be made with clean Diesel fuel which has a

Fig. IH359—IH injection pump drive gear and timing pointer installation.

A. Chisel marked   B. Cap screws
   groove   C. Hub
          D. Timing pointer

viscosity of 35-40 Saybolt Universal seconds at 100° F.

Primary pump pressure.... 45-50 psi
Scavenging valve opening
   pressure .... 20-25 psi
Fuel return check valve
   opening pressure .... 62-68 psi
Reverse check valve open-
   ing pressure .... 450-600 psi
Fuel delivery per nozzle @ 550 rpm
   for one minute.
Series M & 6 Pump No.
   265401 R91 .... 67-71 cc
   65306 DA1 & DA101..... 51.4-54.6 cc
   65306 DA201 .... 66-70 cc
Series 9 Pump No.
   65309 DA2 & DA102.... 61.1-64.9 cc
   65309 DA202 & DA205. 63.05-66.95 cc
   65309 DA5 & DA105..... 61.1-64.9 cc
   268302 R91 .... 76-82 cc
   268305 R91 .... 76-82 cc

The check valve in the check and reverse check valve assembly should have no leakage at 400 psi for one minute.

Distributor valve-to-seat leakage at 2500 psi—1 drop per min. per valve.
Distributor valve-to-bushing-to-gasket leakage at 2500 psi—½ cc per min. per valve.

Leakage past the stem of any distributor valve should not exceed ½ cc per min. at 2500 psi.

132B. To test the condition of the primary pump, when the fuel filters and water trap are known to be in good condition (refer to paragraph 123), start engine, run on gasoline cycle and observe pressure gage reading as shown on the gage which is attached to the top of the pump. If hand is in the "RED", renew or recondition the pump as outlined in paragraph 135 or 136.

To test the operation of the scavenging valve, obtain a spare scavenging valve cap and drill a ⅛ inch hole in the center of the screwdriver slot. Install the drilled cap in place of the regular scavenging valve cap (73—Fig. IH356). Start engine and insert a small diameter rod in the ⅛ inch hole and against the scavenging valve. If the valve is operating properly, it will "kick" the rod approximately ¼ inch. If the valve action is sluggish, remove, clean, and/or renew the valve as outlined in paragraph 137 or 138.

To test the condition of the distributor valve, remove the nozzle tubes from the injection pump discharge fittings, start engine and run

35

on the gasoline cycle. Advance the engine speed control lever and observe the fuel flow from the discharge fittings. Excessive amounts of fuel (considerably more fuel from one fitting than the others) being discharged from any one fitting indicates failure within the distributor block—a defective distributor valve will pass fuel on each injection rather than each fourth injection of the pump.

If engine dies at low speeds, loses power and/or surges, and the governor, fuel filters and primary pump are known to be in good condition, it is reasonable to suspect a faulty pump plunger unit.

133. INJECTION PUMP—R & R. To remove the complete injection pump unit, first shut off the fuel supply and thoroughly clean dirt from pump, fuel lines and connections. Disconnect fuel lines, breather pipe, nozzle pipes and speed control rod from pump, and remove glass bowl from water trap. Remove the pump gear cover from timing gear case and remove the timing pointer by taking out the three cap screws (B—Fig. IH359) which hold the pointer, gear and hub together. Remove the cap screws which retain pump to crankcase front cover and remove the pump.

Note: Three of the pump retaining cap screws are located inside the timing gear housing and must be removed through the openings in the pump drive gear.

To reinstall the injection pump, crank engine until No. 1 piston is coming up on its compression stroke and the notch marked "DC" on the crankshaft pulley is in register with pointer on the crankcase front cover. Install pump. Turn the injection pump gear hub (C) so that the chisel marked groove (A) on the gear hub is in line with the zero degree graduation on the drive gear as shown. Install the timing indicator with pointer at the 0° position and bolt the gear hub, gear and indicator together in this position.

Recheck the injection timing as outlined in paragraph 130A.

134. OVERHAUL. Unless complete pump testing equipment is available, there is little actual overhaul work which can be accomplished on the injection pump; however, if the engine does not run properly, or not at all, and the checks outlined in paragraph 132 point to a faulty injection pump, the complete injection pump can be renewed or some of the component units of the pump can be renewed **without** the use of special testing

equipment. The procedure for R&R of the component units is given in paragraphs 135, 137, 139 and 141.

If complete pump testing equipment is available, the component units of the injection pump can be disassembled and overhauled, using the specifications which are listed in paragraphs 136, 138, 140 and 142.

135. PRIMARY PUMP—R & R AS A UNIT. To remove the primary pump (which includes the scavenging pump unit), as shown in Fig. IH361, first remove glass bowl from water trap and disconnect the fuel lines from the pump. Remove the two cap screws which retain the primary pump to the injection pump and remove the pump.

Install the primary pump by reversing the removal procedure, making certain that the holes in gasket (2) are properly aligned with holes in the injection pump housing. Bleed the fuel system as outlined in paragraph 124.

136. PRIMARY PUMP—DISASSEMBLE AND OVERHAUL. To disassemble the pump, refer to Fig. IH362 and proceed as outlined in the following steps: Remove housing cover (5), gasket (4), by-pass valve spring (9) and by-pass valve (10). Turn the pump drive shaft slightly and remove the primary pump idler gear. Push in on the pump shaft coupling, push in on the primary pump drive gear and remove snap ring (22), primary pump drive gear and the drive gear Woodruff key. Remove the three screws which retain the bearing cage (16) to the pump housing, remove the bearing cage assembly and withdraw the scavenging pump idler gear and shaft unit. To disassemble the bearing cage assembly, remove the scavenging pump drive gear (15), Woodruff key and withdraw the pump drive shaft (21). Remove nut (20) and withdraw spring (19), gland (18) and packings (17) from the bearing cage.

Fig. IH361—Type A primary and scavenging pump unit (1) removed from the injection pump. The scavenging pump is not on type B injection pump.

2. Bearing cage gasket
3. Housing gasket

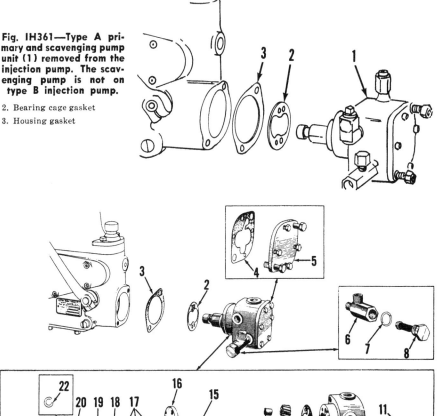

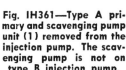

Fig. IH362—Exploded view of type A primary and scavenging pump unit. On type B injection pumps, the scavenging pump is not used.

2 & 3. Gasket	9. By-pass valve spring	13. Scavenging pump idler gear
4. Cover gasket	10. By-pass valve	14. Bushing
5. Cover	11. Primary pump gears	15. Scavenging pump drive gear
6. Filter body	12. Gasket	16. **Bearing cage**
7. Filter gasket		17. Packing assembly
8. Filter		18. Packing gland
		19. Spring
		20. Nut
		21. Pump shaft
		22. Snap ring

Note: The pump drive shaft and idler gear shaft are carried in two bushings each in the pump housing and the drive shaft is carried in one bushing in the bearing cage. These bushings can be renewed if excessively worn.

Thoroughly clean all parts in a good solvent and check the pump parts against the values listed below:

Thickness of housing gasket.1/16 inch
Thickness of housing cover
     gasket ....0.004 or 0.006
Thickness of bearing cage
     gasket ....1/32 inch
Counterbore depth in hous-
     ing cover 0.100-0.105
By-pass valve spring free
     length ....1⅜ inches
     Test load and length........
           ......29.8-33 oz. @ ¾ inch
By-pass valve face angle.....60 deg.
By-pass valve seat angle.....64 deg.
By-pass valve seat width.......0.005
Clearance between the
     primary pump idler
     gear and bore in pump
     housing 0.0015-0.0025
Primary pump idler gear
     shaft bore 0.3123-0.3128
Backlash between gears....0.002-0.004
Primary pump idler gear
     thickness...........0.5020-0.5025
Drive gears fit on shaft.0.0003-0.0009L
Drive shaft clearance in
     bushing 0.0005-0.0013
Clearance between the
     scavenging pump idler
     gear and bore in pump
     housing 0.002-0.0045
Clearance between pump
     shaft and packing
     gland 0.005-0.0083
Packing spring free length.17/32 inch
     Test load and
     length ....4½ lbs. @ 5/16 inch
Bushing I.D. 0.3125-0.3130
Gears end play..........0.0015-0.003

Note: The by-pass valve maintains a pressure of 35-40 psi on the final fuel filter line. If the injection pump pressure gage was in the "RED" and fuel filters and primary pump are known to be in good condition, install a new primary pump housing, bearing cage and by-pass valve assembly.

Slight wear on the face (side facing gears) of the pump housing cover and bearing cage can be corrected by lapping.

Reassemble the pump by reversing the disassembly procedure.

Binding of the pump shaft is normal on a reconditioned pump where the shaft packing has been renewed. Due to this binding condition, it is advis- able to run the pump on a test stand for at least ½ hour with complete fuel connections to insure proper lubrication.

137. SCAVENGING VALVE—R & R. To remove the scavenging valve (A—Fig. IH364), remove the scavenging valve cap (13) and withdraw the scavenging valve body. It may be necessary to use IH service tool SE-1330-1 to remove the body. Using this tool remove the scavenging valve tappet and guide (8) and withdraw screen (7). Drain lubricating oil from the injection pump sump, remove the pump housing bottom plate and remove the scavenging valve strainer (6—Figs. IH364 and 365). Remove the scavenging check valve (B—Fig. IH364).

Note: Before installing a complete new scavenging valve, it is advisable to thoroughly clean the valve, reinstall same by reversing the removal procedure and check the valve action. Often a thorough cleaning job will restore the valve to its original operating efficiency.

138. SCAVENGING VALVE—OVERHAUL. Thoroughly clean all parts and examine same for being excessively worn. The scavenging valve tappet has a normal clearance of 0.0004 in the tappet guide.

The tappet-to-guide clearance is O.K. if the tappet will slide *slowly* through the guide when the tappet is propelled by its own weight. The tappet and guide are available as a matched unit only and should be renewed if the clearance is excessive.

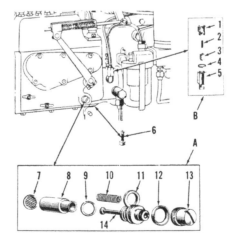

Fig. IH364—Exploded view of type A scavenging valve (A) and scavenging check valve (B).

1. Fitting
2. Check valve spring
3. Check valve
4. Gasket
5. Scavenging check valve body
6. Scavenging strainer
7. Screen
8. Scavenging valve tappet
9. Gasket
10. Valve spring
11. Body seal
12. Cap gasket
13. Cap
14. Valve body

Check the scavenging valve spring against the values listed below.

Free length ....1 9/64 inches
Test load @ length..29 oz. @ ¾ inch

Renew the spring if it is rusted, discolored or does not meet the foregoing specifications.

139. PLUNGER—R&R AS A UNIT. To remove the plunger unit (D—Fig. IH366), first clean dirt from the injection pump, fuel lines and connections and remove the pressure gage from the fuel inlet fitting. Remove the fuel return pipe and washers and remove the injection pump oil filler pipe. Remove the nut from each injection pump discharge fitting and remove the housing cover. Remove the high pressure pipe (E), cap screws (F) and lift the plunger unit from the injection pump.

Note: Before installing a complete new plunger unit, it is advisable to thoroughly clean the plunger unit, reinstall same and check the pump action. Often a thorough cleaning job will restore the plunger to its original operating efficiency.

To install the plunger unit, turn the control gear so that the double-width tooth space is centered in the cutaway portion of the plunger bushing retainer and insert plunger unit so that the double-width tooth of the rack will engage the double-width tooth space of the gear. See Fig. IH368. Install the plunger retaining cap screws and check for binding by moving the rack through full range of travel. Slight binding condition can be corrected by shifting the plunger in its mounting. Install the high pressure line and tighten the retaining cap screws to a torque of 20 ft.-lbs. Install the remaining parts by reversing the removal procedure.

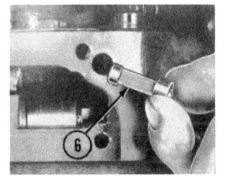

Fig. IH365—Bottom view of type A injection pump with bottom plate removed. The scavenging valve strainer (6) can be removed for cleaning purposes.

140. PLUNGER UNIT—OVERHAUL. To overhaul the plunger unit, refer to Fig. IH369. Completely disassemble the unit and clean all parts in an approved solvent. Check and renew any parts which do not meet the following specifications.

Plunger spring free length . 2.00 inches
Test load and
length . . . . . . 22 lbs. @ 1 9/16 inches
Backlash between control
gear and rack 0.002-0.004
Fuel return check valve
opening pressure . . . . 62-68 psi
Check valve
leakage . . . . none under 400 psi
Reverse check valve
opening pressure . . . . . . 450-600 psi
Fuel return check spring
free length . . . . 25/32 inch
Test load and
length . . 47.8-52.8 oz. @ 35/64 inch

141. DISTRIBUTOR BLOCK—R & R AS A UNIT. To remove the distributor block unit (G—Fig. IH366), remove the pump pressure gage, fuel return pipe and the injection pump oil filler pipe. Remove the nut from each injection pump discharge fitting and remove the housing cover. Remove the high pressure pipe (E) and discharge fitting locks (H). Using a special tool, remove the four discharge fittings (J). Remove the distributor block retaining cap screws and lift the distributor block from the pump housing.

Note: Before installing a complete new distributor block, it is advisable to thoroughly clean the unit, reinstall same and check the pump action. Of-

ten a thorough cleaning job will restore the distributor block to its original operating efficiency.

Install the distributor block by reversing the removal procedure and tighten the distributor block and high pressure line cap screws to a torque of 20 ft.-lbs. Tighten the discharge fittings to a torque of 30 ft.-lbs.

142. DISTRIBUTOR BLOCK—OVERHAUL. To overhaul the distributor block, refer to Fig. IH370, com-

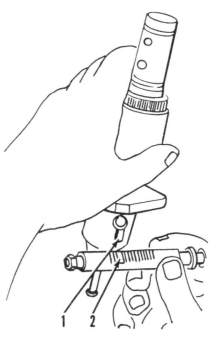

Fig. IH368—Install IH plunger unit so that double width tooth space (1) in gear meshes with double width tooth (2) on rack.

pletely disassemble the unit and clean all parts in an approved solvent. Check and renew any parts which do not meet the following specifications.
*Distributor valve standout
below bottom face of distributor block 0.045-0.050
Distributor valve lift 0.012-0.019
Valve spring test
load and length . 16 lbs. @ 7/16 inch
Distributor valve-to-
seat leakage at
2500 psi . . 1 drop per min. per valve
Distributor valve-to-bushing-
to-gasket leakage at 2500
psi . . . . ½ cc per min. per valve
*Obtained by varying the number of gaskets (3).

143. CAMSHAFT—R & R AND OVERHAUL. To remove the camshaft, first remove the distributor block unit as in paragraph 141, and proceed as follows: Remove hub and oil sling-

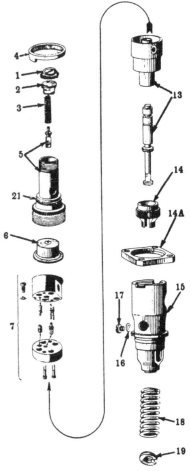

Fig. IH369—Exploded view of IH plunger unit.

1. Fuel return check spring lock
2. Check spring seat
3. Check spring
4. Plunger bushing clamp lock
5. By-pass valve and plunger bushing clamp
6. Filter element
7. Check & reverse check block
13. Plunger & bushing
14. Control gear
14A. Flange
15. Bushing retainer
16. Gasket
17. Screw
18. Plunger spring
19. Spring seat

Fig. IH366—Side view of IH injection pump with the housing cover removed.

D. Plunger unit
E. High pressure pipe
F. Cap screws
G. Distributor block
H. Locks
J. Discharge fitting
X. Index notches
50. Lever
51. Shaft

er unit (7—Fig. IH372). Remove the governor shaft thrust screws (Fig. IH373) from the pump mounting flange. Remove the mounting flange from the pump housing and remove the drive hub Woodruff key from the camshaft. Remove the nut which retains the governor drive gear to the governor shaft and using a suitable puller, remove the camshaft gear. Remove the camshaft front bearing retainer and withdraw camshaft from pump housing.

Note: It may be necessary to reinstall the drive hub retaining nut and bump the camshaft out by tapping the nut with a soft hammer. If the camshaft rear bearing remains in the pump housing, it can be removed by using a special puller.

Remove the distributor valve tappets and identify them with respect to the bores from which they were removed.

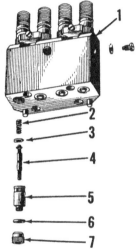

Fig. IH370—Partially exploded view of IH injection pump distributor block.

1. Distributor block assy.	4. Valve
2. Valve spring	5. Valve bushing
3. Gasket (0.031)	6. Gasket
	7. Bushing clamp

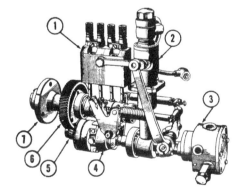

Fig. IH372—Skeleton view of IH injection pump, showing the relation of the various component parts.

1. Distributor block	5. Drive gear
2. Plunger unit	6. Camshaft gear
3. Primary pump	7. Hub and oil
4. Governor unit	slinger

Inspect all parts for damage or for being excessively worn and renew any which do not meet the following specifications.

Camshaft lobe height..........0.020
Maximum allowable wear......0.0035
Tappet diameter 0.2491-0.2493
Tappets clearance in
   bores 0.0004-0.0009
Tappet standout above
   pump housing when camshaft lobes are in lowest
   position 0.052-0.054

Reinstall the camshaft by reversing the removal procedure and tighten the camshaft front bearing retainer cap screws to a torque of 7½-10 ft.-lbs. Install camshaft gear with shallow

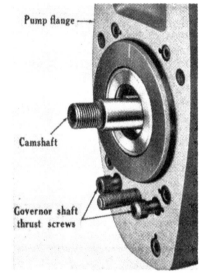

Fig. IH373—Front view of IH injection pump with hub and oil slinger unit removed.

Fig. IH375—When installing the IH injection pump camshaft and/or governor gears, mesh the single beveled tooth on the camshaft gear with the two beveled teeth on the governor gear as shown in white circle.

hub toward pump housing and mesh single beveled tooth on camshaft gear between the two beveled teeth on the governor drive gear as shown in Fig. IH375. Tighten the governor drive gear retaining nut to a torque of 45 ft.-lbs. Install the pump mounting flange and tighten the governor shaft thrust screws to a torque of 5 ft.-lbs. Install the pump drive hub and tighten the hub retaining nut to a torque of 115-125 ft.-lbs.

144. GOVERNOR—ADJUST. To adjust the governor, first remove the pump housing oil filler neck and loosen lock nut (1—Fig. IH377) on the rack. Remove the pump housing side cover and adjust the torque lever stop screw (10—Fig. IH378) until the face of the screw which contacts the torque lever stop pin (11) is dimension A-Table X from the finished surface of the torque lever (12). Adjust the torque shoe screw (13) so the lower face of the shoe is dimension B from finished surface beneath the lock nut. Adjust the torque lever pick up screw (14) to give dimension (C) between the torque lever stop screw (10) and the torque lever stop pin (11). With the speed control lever in full shut-off position, adjust stop screw (2—Fig. IH377) until there is $\frac{1}{16}$-$\frac{1}{8}$ inch clearance between the pump housing at point (X—Fig. IH378) and the forward end of the torque control assembly.

Remove one injection nozzle and reconnect the nozzle to the nozzle tube. With the speed control rod in closed position, start engine and run on gasoline cycle. Turn the rack adjusting nut (3—Fig. IH377) until injection just starts. Then turn the nut back until injection just stops. To insure positive shut-off, turn the nut 1½ additional turns in the same direction

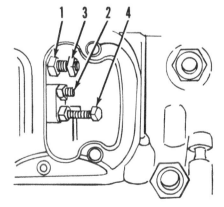

Fig. IH377—Top view of IH injection pump with oil filler neck removed.

1. Lock nut	3. Adjusting nut
2. Shut-off stop screw	4. High speed stop screw

(direction which stopped injection). Adjust the engine high idle speed (E) with stop screw (4).

With engine running at high idle speed, adjust the gap between torque arm stop screw (15—Fig. IH378) and torque arm (16) to dimension (F) by turning stop screw (15). Adjust the bumper spring stop (17) to give a slight pressure on the bumper spring (18) at high idle speed. If the engine surges, increase the bumper spring pressure slightly. Recheck the torque lever pick up screw adjustment, high idle speed and make certain that engine will shut off on the Diesel cycle.

Note: The tractor should now be ready to test under actual load conditions (as if working in the field).

If engine surges at high idle speed, increase the pressure on the bumper spring by adjusting the bumper spring stop.

If engine lacks power at rated load speed and overload speed, increase the gap between the torque arm (16—Fig. IH378) and the torque arm stop screw (15) by backing out the screw. Do not adjust the stop screw to a point where exhaust smoke is excessive. Conversely, if smoke and power are both excessive, decrease the gap.

If engine tends to stall easily and lacks power at overload speeds, back out the torque lever pickup screw (14). Do not adjust the screw to a point where exhaust smoke is excessive.

If engine performance is satisfactory except for excessive smoke at overload speeds, turn the torque lever pickup screw in slightly.

If tractor seems to lack power at rated load speed, but power is O.K. at overload speeds, adjust the high speed stop screw (4—Fig. IH377) to permit pulling the control lever farther back.

Table X
Diesel Engine Governor Adjustments

A. Torque Lever Stop Screw
MD-MDV-WD6-OSD6 ......$\frac{3}{8}$ inch
Super M & 6-Series 9.......$\frac{7}{16}$ inch
B. Torque Lever Shoe Screw
Series M & 6.............$\frac{5}{8}$ inch
WD9-WDR9 ....$\frac{11}{16}$ inch
Super WD9-Super
WDR9 ....41/64 inch
C. Overload Gap
MD-MDV-WD6-OSD6 ......$\frac{1}{16}$ inch
Super M & 6.............0.030
WD9-WDR9 ....$\frac{1}{16}$ inch
Super WD9-Super
WDR9 ....0.032-0.072
E. High Idle Speed (±30 rpm)
MD-MDV-WD6-OSD6 1610
Super M & 6.............1580
WD9-WDR9 1665
Super WD9-Super WDR9.....1635
F. High Idle Gap
MD-MDV-WD6-OSD6 ....0.085
Super M & 6.............0.060
WD9-WDR9 ....0.090
Super WD9-Super WDR9.....0.108

145. GOVERNOR UNIT—R&R AND OVERHAUL. To remove the governor unit, first remove the primary pump, plunger unit, distributor block unit, camshaft unit and proceed as follows: Remove the speed control lever, pump housing side cover, and torque control unit. Remove the pump housing back cover and withdraw the governor fork shaft. Remove the rack locating screw from back of pump housing and push the rack forward and free of the governor fork, then pull the rack rearward. Using a small

brass rod inserted through rear of pump housing, drive the governor shaft assembly from the pump housing, and remove the governor fork. Remove the expansion plug from rear of pump housing and withdraw the rack. The plunger tappet can be removed at this time. The need and procedure for further disassembly is evident after an examination of the unit.

Inspect all parts for damage or for being excessively worn and renew any which do not meet the following specifications.

Weight pins fit in
carrier 0.0003-0.0009
Weight pin diameter.....0.2814-0.2817
Clearance between fork shaft
spacer and bearing....0.001-0.0015
Clearance of thrust shoe pins in
thrust shoes 0.001-0.0015
Backlash between rack and control
gear 0.002-0.004
Distance of top of plunger tappet
bushing below top of pump housing ....1⅝ inches
Backlash between governor gear
and camshaft gear...........0.004
Clearance between governor sleeve
and governor shaft....0.0008-0.0038
Eccentric bearing sleeve end
play 0.010-0.025
Distance of inner end of pump
housing bushing to rear face of
pump housing........2 9/64 inches
I. D. of pump housing bushing 0.406-0.407
Distance of control lever shaft oil
seal below outside face of pump
housing ....1/16 inch

When reassembling the governor, tighten the shaft nuts to a torque of 40-45 ft.-lbs. Install governor by reversing the removal procedure.

When installing the torque lever assembly, hold the assembly to the rear as far as possible and tighten the assembly retaining nut to a torque of 8-10 ft.-lbs. When installing the speed control lever, make certain that notch in lever (50—Fig. IH366) is in register with notch in the lever shaft (51) as shown.

Adjust the governor unit as outlined in paragraph 144.

DIESEL STARTING SYSTEM
Series M-6-9

International Diesel engines are started on ordinary gasoline after first closing the Diesel engine throttle and pulling the compression release lever. Pulling the compression release lever accomplishes four things: A starting valve in each engine cylinder is opened, thereby enlarging the combustion chamber and reducing the compression ratio to approximately

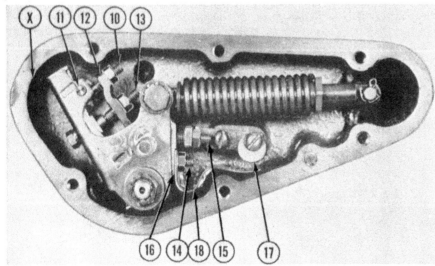

Fig. IH378—Side view of IH injection pump with the pump housing side cover removed

10. Torque lever stop screw	12. Torque lever	15. Arm stop screw
11. Torque lever stop pin	13. Torque shoe screw	16. Torque arm
	14. Torque lever pick up screw	17. Bumper spring stop
		18. Bumper spring

6.4 to 1. It closes two butterfly valves in the Diesel air intake manifold, and allows air to pass through the starting carburetor. The magneto or battery ignition electrical circuit is completed by opening of the grounding switch which is located in the forward portion of the intake manifold. It releases the float in the carburetor, allowing the carburetor inlet needle to be actuated by the float.

After the engine runs on gasoline approximately one minute, the engine is switched over to full Diesel operation by releasing the compression release mechanism and at the same time opening the Diesel engine throttle. Releasing the compression release mechanism opens the manifold butterfly valves, closes the cylinder head starting valves, locks the carburetor needle valve on its seat and closes the magneto or battery ignition grounding switch.

150. **ADJUSTMENTS.** If the starting controls have been removed, or if erratic action occurs, inspect all parts for wear, renew any which are questionable and adjust the mechanism as follows:

Note: Before attempting to adjust the starting mechanism, remove the valve cover and complete manifold from cylinder head and disconnect yoke (F—Fig. IH382) from lever (V).

First step in the adjustment procedure is to set the starting controls for Diesel operation. Adjust screw (R—Fig. IH382) to obtain 0.060 clearance between jaw on (B) and latch (A). Adjust yoke (F) on rod (G) so that (when yoke is connected) there will be 0.060-0.080 clearance between each starting valve shaft cam (H) and the valve cover (I). If the clearance cannot be obtained on all starting valves, check for excessively worn cams (H) or for a twisted starting valve shaft. Rotate the cross shaft with the com-

pression release lever until lever (E) contacts pick up face of jaw on (B) at point (X), then adjust set screw (C) on bracket (T) to give 0.100 clearance between set screw (C) and lever (D).

Second step in the adjustment procedure is to set the starting controls in the gasoline starting position. Adjust set screw (U) in jaw (B) to obtain 0.015 clearance between set screw (U) and bracket (S) at point (UU). Now assemble the manifold to the engine and connect yoke (F).

Third step in the adjustment procedure is to start the engine and run on the Diesel cycle. Retard the engine speed control lever until the poppet in lever (L) locates in the hole in the stationary disc and adjust nuts (N) on rod (W) until springs (P) and (Q) are equal in length and the recommended low idle speed (listed below) is obtained.
Super M-6-9 Series.......450-550 rpm
Other Models ....425 rpm

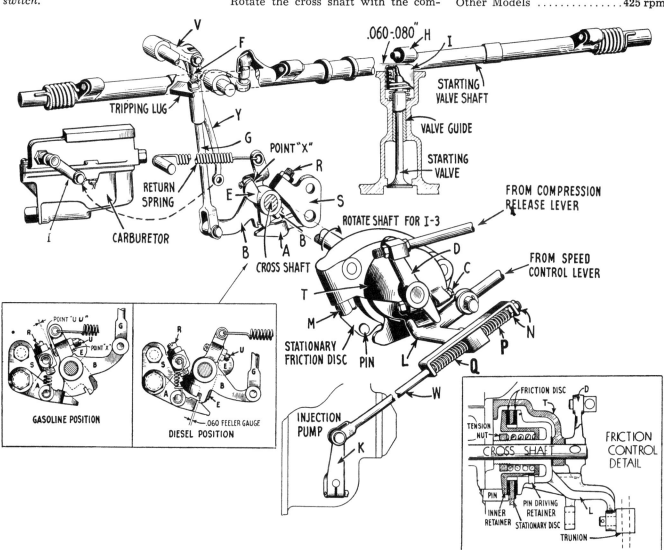

Fig. IH382—IH Diesel engine starting system linkage, showing governor friction control and associated parts.

Move the speed control lever to off position and adjust the yoke at rear end of speed control rod until spring (P) is partially compressed and the flat spot on lever (L) contacts bracket (T). Move speed control lever to full throttle position and adjust the stop at rear of the speed control rod until spring (Q) is partially compressed.

151. GOVERNOR FRICTION CONTROL. The governor friction control which serves to hold the throttle at selected positions, is located on the left end of the starting mechanism cross shaft as shown in Fig. IH382. To adjust the spring tension of the friction control mechanism, remove the unit from tractor, reverse the unit and temporarily reinstall the unit (backwards) on the cross shaft. Tighten the tension nut until a pull of 70-75 lbs. is obtained as shown in Fig. IH384. Then, reinstall in correct position.

152. INTAKE MANIFOLD. The battery ignition or magneto grounding switch and the two butterfly valves which close off the Diesel air intake passages during the gasoline cycle are located in the intake manifold. The linkage which is shown exploded from the manifold in Fig. IH386 must be free with absolutely no tendency to bind. Adjust the butterfly valves (4) with set screws (A) so that valves are perfectly horizontal inside the air pas-

sages when the levers on the shafts are against the set screws. Bend the prong (if necessary) on the cut-out terminal plate to assure good contact

of the grounding switch. There should be no spark at the spark plugs when engine is running on the Diesel cycle.

Fig. IH384 — Checking spring tension of the governor friction control. Notice that the friction control unit is installed backwards for this check.

A. Set screw
1. Exhaust manifold
2. Exhaust pipe
3. Seals
4. Butterfly valve
5. Front control shaft
6. Gasket
7. End cover
8. Control spring
9. Spring stud
10. Switch
11. Intake manifold
12. Rear control shaft
13. Control lever

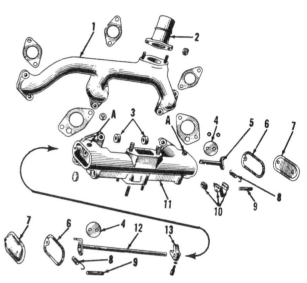

Fig. IH386—Exploded view of typical IH Diesel engine manifold and associated parts.

NON-DIESEL GOVERNOR

International Harvester governors are the centrifugal flyweight type and are driven by a gear in the timing gear train. Before attempting any governor adjustments, check the operating linkage for binding condition or lost motion and correct any undesirable conditions.

Series A-B-C

155. ADJUSTMENT. To adjust the governor (engine stopped), place the speed change lever in wide open position and remove the clevis pin (12—Fig. IH390) from the governor rockshaft arm. Hold the rockshaft arm (13) and carburetor throttle rod (10) as far toward carburetor as they will go; at which time, pin (12) should slide freely into place. If the pin will not slide freely into place, adjust the length of rod (10) until proper register is obtained.

High idle speed adjustment is made with speed change lever in wide open position by turning adjusting screw (8) which is located on top of governor housing.

Hunting or unsteady running can be eliminated by removing cap (22) and turning the bumper spring screw (21) *in*. Do not turn the bumper spring screw in too far as it will interfere with the low idle speed adjustment. CAUTION: Stop the engine before making the bumper spring adjustment.

Governed Speeds
Low Idle
    Series A-B ....525 rpm
    Series C ....425 rpm
High Idle (No Load)
    Series A-B ....1540 rpm
    Series C ....1875 rpm
Belt Pulley (No Load)
    Series A-B ....1272 rpm
    Series C ....1549 rpm
156. R&R AND OVERHAUL. Remove the magneto or distributor and mark location of slots in magneto drive coupling on governor gear (1—Fig. IH390) in relation to crankcase. This procedure will facilitate reinstallation of governor gear in proper timing mesh with camshaft gear if position of crankshaft is not changed.

Remove governor speed control rod then remove the governor assembly from timing gear cover.

When disassembling, do not bend hooks of governor spring (5). The spring lever (4) should be removed from the speed change lever shaft to remove and install the spring. Upper hook of spring should be inserted in lever shaft so that open end of hook will face the side of governor housing. Place the lower hook of spring in hole (A) of rockshaft lever.

Two needle bearings (6) and one "Oilite" bushing (15) which support the rockshaft (16) can be renewed when worn. The felt seal (14) which is assembled outside the bushing should not place any drag on the rockshaft.

The bushing in the crankcase which supports the governor and magneto or distributor gear hub may be renewed when governor and magneto are out. The I&T recommended clearance of gear hub in bushing is 0.0015 to 0.002. Refer to the following Table for additional governor data.

Governor Spring Data

Free length (inside of hook to
inside of hook) ....1 15/16 inches
Test load and
length ....15¼ lbs. @ 2⅛ inches

Cub

158. **ADJUSTMENT.** To adjust the governor (engine stopped), place the speed change lever in wide open position and remove pin (12—Fig. IH391) from rockshaft arm (13). Hold the rockshaft arm (13) and carburetor throttle rod (10) as far toward carburetor as they will go; at which time, pin (12) should slide into place. If the pin will not slide freely into place, adjust the length of rod (10) until proper register is obtained.

High idle speed adjustment is made with speed change lever in wide open position by turning adjusting screw (8). Turn screw (8) *in* to decrease speed and *out* to increase speed.

Hunting or unsteady running can be eliminated by the bumper spring adjusting screw (21) which is located at front of the governor housing. Do not turn the bumper spring screw in too far as it will interfere with the low idle speed adjustment.

Governed Speeds

Low Idle .... 500 rpm
High Idle (No Load) ....1800 rpm
Belt Pulley (No Load)......1487 rpm

159. **R&R AND OVERHAUL.** Crank engine until No. 1 (front) piston is on compression stroke, then crank slowly until the DC notch on crankshaft pulley is in register with pointer on crankcase front cover; remove magneto or distributor (it may be noted that there is a mark on crankcase which registers with magneto drive slots on magneto or distributor governor gear. Use this mark to retime gear to idler gear when reinstalling governor). Governor assembly can now be removed.

Disassembly is apparent after studying the unit and Fig. IH391. The front bushing (28) and rear bushing (27) are renewable. The rear race of the thrust bearing is pressed on the thrust sleeve (2) but the front or outer race should rotate freely on the sleeve. End play of governor shaft should be 0.004-0.013. Refer to the following Table for additional governor data.

Governor Spring Data

Free length (inside of hook to
inside of hook) ....1½ inches
Test load and
length ....28½ lbs. @ 1⅞ inches

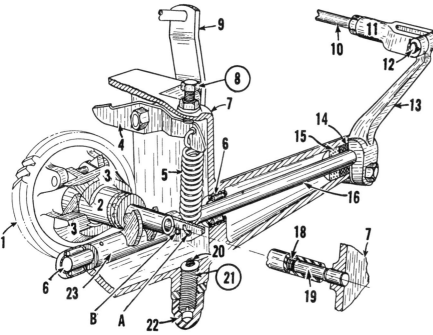

Fig. IH390—Cut away view of series A-B-C governor unit. The governor is driven from the timing gear train.

1. Magneto-governor gear	7. Governor housing	12. Clevis pin
2. Sleeve & thrust bearing	8. Speed adjusting screw	13. Rockshaft arm
3. Governor weights	9. Speed change lever	14. Felt seal
4. Spring lever	10. Throttle rod	15. Oilite bushing
5. Governor spring	11. Throttle rod clevis (yoke)	16. Rockshaft
6. Needle bearings		18. Thrust spring
		19. Thrust pin
		20. Surge (bumper) spring
		21. Surge spring adjusting screw
		22. Spring body cap
		23. Rockshaft lever
		A-B. Spring position holes

Fig. IH391—Sectional view of Cub governor. Shaft end play should be 0.004-0.013 as shown.

1. Magneto-governor gear
2. Governor sleeve
3. **Governor weights**
5. Governor spring
6. Rockshaft needle bearing
8. **Speed adjusting screw**
9. Speed change lever
10. Throttle rod
12. Connecting pin
13. Rockshaft arm
14. Rockshaft oil seal
16. Rockshaft
17. Governor housing
20. Surge (bumper) spring
21. Spring adjusting screw
23. Rockshaft fork
24. Weight hinge pin
25. Rockshaft bracket
26. Rockshaft extension
27. Rear bushing
28. Front bushing
29. Governor base
30. Sleeve stop (snap ring)

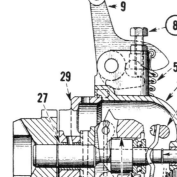
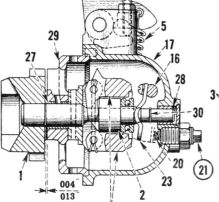
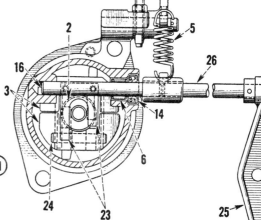

When assembling fork (23) to rockshaft (16) the fingers of fork must clear the sleeve (2). Install rockshaft oil seal (14) with lip toward needle bearing (6). Be sure governor sleeve stop ring (30) and bumper spring (20) are in place. Before installing governor to engine, check condition of the governor-magneto or distributor gear seal in the crankcase and renew if necessary. Seal is installed with lip toward gear and is pressed into bore to a distance of 3/16 inch measured from front face of seal to front of crankcase. Seal must be installed evenly and can be checked by measuring through rear opening of bore. (Use shellac or similar sealer on outside diameter of seal when installing).

When reinstalling, mesh magneto or distributor-governor gear with idler gear so that mark on the gear is in register with the mark on crankcase as previously explained. Install magneto or distributor.

Series H-4 & Non Diesel M-6-9

161. **ADJUST.** Adjust the carburetor; then with engine stopped, remove cover from side of governor housing. Turn the governor screw (10—Fig. IH393) until it just touches its stop (there should be no tension on governor spring (2)). Place operator's hand lever in full speed position and adjust high idle speed screw (11) until it just touches stop (12) (if length of threads on screw (11) will not permit this adjustment, lengthen or shorten the linkage between operator's hand lever and governor lever (3) until the screw is brought into range).

To adjust governor to speeds which follow, start engine and place operator's hand lever in full speed position; turn high idle speed screw (11) *in* to decrease speed or *out* to increase speed.

If correct speed adjustment cannot be obtained by the above procedure or if governor is just being installed on engine, adjust the governor to carburetor linkage as follows: With engine stopped, place operator's hand lever in full speed position and remove cover from top slanted surface of connecting rod housing (22—Fig. IH394). Remove the pin (18) from the throttle shaft lever (23) and adjusting block (19); pull both the throttle

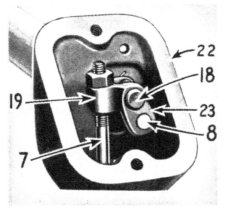

Fig. IH394—Synchronizing governor to carburetor linkage on series H and 4 and non-Diesels of the M, 6 and 9 series.

7. Connecting rod
8. Throttle shaft
18. Connecting rod pin
19. Adjusting block
22. Connecting rod housing
23. Throttle shaft lever

shaft lever and the adjusting block up as far as they will go. With the parts thus held, it should be possible freely to insert pin (18). If pin does not enter freely, screw the block (19) on or off connecting rod (7) until this condition is obtained. The carburetor throttle butterfly is now synchronized to the governor, with governor in wide open position.

Governed Speeds
Low Idle
    Super H & 4 Series.........425 rpm
    Other Series H & 4.........450 rpm
    Series M-6-9 ....425 rpm

High Idle (No Load)
    Super H & 4 Series.......1864 rpm
    Other Series H & 4.......1815 rpm
    Super M & 6 Series.......1600 rpm
    Other M & 6 Series.......1595 rpm
    Series 9 ....1650 rpm

Belt Pulley (No Load)
    Series H-4 ....1121 rpm
    Super M & 6 Series.......992 rpm
    Other M & 6 Series.......989 rpm
    Series 9 .... 778 rpm

162. **R&R AND OVERHAUL.** To remove the unit from engine, loosen carburetor mounting screws and remove cap screws which hold front face of governor housing to the engine. Also remove ventilating tube (17—Fig. IH395) and screws (15) from throttle shaft housing retainer. Pull throttle shaft housing (24) forward far enough to release throttle shaft (8) from the coupling at carburetor

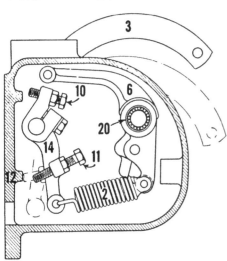

Fig. IH393—Governor adjustments for series H and 4 and non-Diesels of the M, 6 and 9 series.

2. Governor spring
3. Speed change lever
6. Rockshaft lever
10. Idle stop screw
11. Speed adjusting screw
12. Screw stop
14. Governor spring lever
20. Rockshaft lever bearing

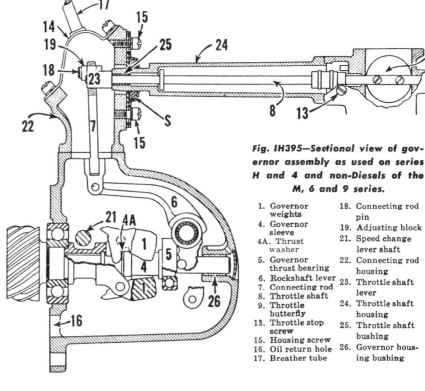

Fig. IH395—Sectional view of governor assembly as used on series H and 4 and non-Diesels of the M, 6 and 9 series.

1. Governor weights
4. Governor sleeve
4A. Thrust washer
5. Governor thrust bearing
6. Rockshaft lever
7. Connecting rod
8. Throttle shaft
9. Throttle butterfly
13. Throttle stop screw
15. Housing screw
16. Oil return hole
17. Breather tube
18. Connecting rod pin
19. Adjusting block
21. Speed change lever shaft
22. Connecting rod housing
23. Throttle shaft lever
24. Throttle shaft housing
25. Throttle shaft bushing
26. Governor housing bushing

butterfly valve. When reinstalling governor unit, be sure throttle shaft fully engages coupling at carburetor. Allow throttle shaft and housing to find correct alignment before tightening screws (15).

The governor shaft forward ball bearing is retained in housing by a snap ring (not shown) which also takes the thrust of the helical drive gear. Check clearance of governor weights (1) on their hinge pins and freedom of governor sleeve (4) on governor shaft. The bushing (26) at rear end of shaft is renewable after removing the Welch plug from body. The ball thrust bearing (5) should be installed on sleeve (4) with the closed face away from the shoulder on sleeve. Needle bearings supporting the governor lever (6) are caged units. Linkage between governor lever (6) and carburetor should operate without binding.

COOLING SYSTEM

Models A, AV, B, BN, Super A & AV, Cub and early production C tractors were factory equipped with a non-pressure, thermo-siphon type cooling system. Late production C tractors were equipped with a thermo-siphon cooling system and a pressure type radiator cap. Model C tractors can be converted to a pump circulating type system by installing "Water Pump Attachment No. 356933R91." The Super C tractor is factory equipped with a water pump and a pressure type system.

Note: *On late production Super A tractors and for replacement purposes on models A, B and early Super A tractors, a pressure type radiator is used. This radiator can be equipped with either a pressure or non-pressure type radiator cap.*

Series H, M, 4, 6 and 9 tractors are factory equipped with a water pump, and all but early production tractors have a pressure type system. A water directing header plate (or plates on Diesel tractors) is fitted to the side of the cylinder block. A reverse pitch, blower-type fan is available as optional equipment for all tractors mentioned in this paragraph except Series 9 and Super models of the H, M, 4 and 6 Series.

RADIATOR

Series A-B-C-Cub

165. To remove the radiator, drain cooling system and remove hood and grille (screen on Cub). Disconnect radiator hoses and on all models except the Cub, disconnect the radiator upper brace rod from the water outlet casting. On series A, B and C remove nuts from both lower radiator mounting bolts and lift radiator assembly from tractor. On the Cub, unbolt radiator from steering gear housing and remove radiator. Note: On the Cub, the lower radiator tank is an integral part of the steering gear housing. On other models, the radiator is a one-piece unit.

Series H-M

166. To remove the radiator, drain cooling system and remove the hood and grille. Disconnect radiator hoses, radiator upper support rod and remove the dust shield from under front of frame rails. Remove the steering worm shaft, unbolt radiator from bolster and remove radiator.

Series 4-6-9

167. To remove the radiator, drain cooling system and remove hood and grille. Disconnect radiator hoses and radiator upper support rod (or brace). Unbolt radiator from main frame and remove radiator.

On late production series 9 tractors, the radiator can be disassembled for cleaning and/or repair.

FAN

Series A-B-C-Cub

168. To remove the fan assembly, first remove radiator from tractor.

Two different fan assemblies (shown in Fig. IH396) have been used on models A, AV, B, BN, Super A, Super AV and C. The complete assembly of either fan can be used; however, the parts are not interchangeable between the two assemblies. The fan assembly shown in the lower view only is used on the Cub and Super C tractors.

Disassembly of either fan is evident after an examination of the unit and reference to Fig. IH396. The fan bearing should have an I&T suggested clearance of 0.0045-0.005 on its spindle.

Series H-M-4-6-9

169. The fan is mounted on and driven by the water pump. Refer to paragraph 175. To remove fan blades it is necessary to first remove the radiator.

THERMOSTAT

All Models

171. The thermostat which is located in the engine water outlet casting is used on all tractors except those equipped with a thermo-siphon type cooling system. Super C tractors and C tractors equipped with a water pump attachment, have a 165-195 degree thermostat. All tractors which are equipped with a non-pressure cooling system, have a 130-155 degree thermostat. All other models have a 165-190 degree thermostat.

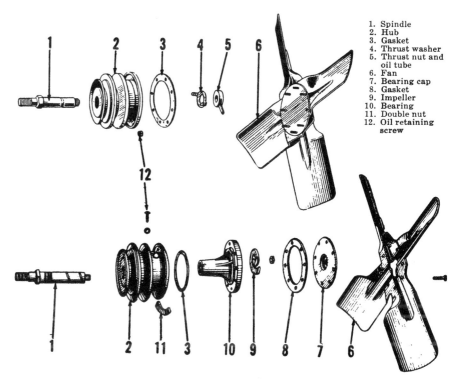

1. Spindle
2. Hub
3. Gasket
4. Thrust washer
5. Thrust nut and oil tube
6. Fan
7. Bearing cap
8. Gasket
9. Impeller
10. Bearing
11. Double nut
12. Oil retaining screw

Fig. IH396—Exploded views of fan assemblies as used on models A, AV, B, BN, Super A, Super AV and C. The fan assembly shown in the lower view only is used on the Cub and Super C.

Note: Some of the above tractors were equipped with a 150-175 degree thermostat. When servicing these tractors, use a 165-190 degree thermostat.

WATER PUMP
Models Super C-C (Some)

173. **R&R AND OVERHAUL.** To remove the water pump, first drain cooling system and remove the water drain pipe. Remove drive belts from generator and water pump. Loosen the lower radiator hose, remove the pump retaining cap screws and lift water pump from tractor.

To disassemble the pump, remove the cap screws which retain rear body (6—Fig. IH398) to body (7) and separate the two bodies. Remove the bearing retainer wire (2) and using a suitable puller, remove pulley (1). Press the shaft, seal and impeller out of pump body, then press the shaft out of the impeller, and remove seal (3).

Note: the shaft and bearing assembly are available as a preassembled unit only.

To assemble the water pump, install the seal (3), insert the shaft and bearing unit and install retaining wire (2). Support the pump drive shaft and press the impeller on the shaft until the clearance between the impeller vanes and rear face of front body is 0.031-0.041. The clearance is shown at (B). Using a new gasket, install rear body to front body. Press the drive pulley on the drive shaft until the distance between rear face of rear body and front face of pulley is 3 21/32 inches. This dimension is shown at (A).

Series H-M-4-6-9

174. **REMOVE AND REINSTALL.** To remove water pump from tractor, first remove radiator as outlined in paragraph 166 or 167.

175. **REPACK.** Packing is furnished in split segments ¼ inch thick. To renew the packing (19—Fig. IH400 or Fig. IH401), it is advisable first to remove the pump unit from the engine as per paragraph 174. Remove driver pin (5) and driver (6). Unscrew packing nut (3) and remove old packing and replace with new. Impeller and shaft (15) may be removed for easy renewal of packing.

176. **OVERHAUL.** Disassemble the unit in the following manner: Remove driver pin (5—Fig. IH400 or Fig. IH 401), driver (6), packing nut (3), bearing retaining nut (7) and the front oil seal (8). The impeller and shaft (15) can now be withdrawn. Support the unit on rear flange of pulley hub (9—Fig. IH401) or fixed flange (10—Fig. IH400) and press on forward end of sleeve (14) which will release rear bearing (12) and pump body (16) from the fan hub.

On the pump shown in Fig. IH401, the adjustable pulley flange (1) can be removed by unscrewing after first removing the set screw (2). On the

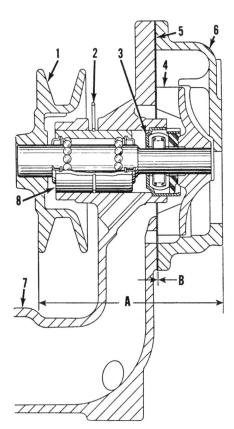

1. Pulley flange
2. Flange set screw
3. Packing nut
4. Bearing lock sleeve
5. Driver pin
6. Pump shaft driver
7. Bearing retaining nut
8. Oil seal
9. Fan hub
10. Fixed flange
11. Flange retaining pin
12. Fan hub bearing
13. Rear oil seal
14. Pump shaft sleeve
15. Shaft with impeller and thrust washer
16. Pump body
17. Felt washer
18. Bearing spacer
19. Pump packing
20. Pump shaft bushings
21. Bearing retainer

Fig. IH398—Sectional view of Super C water pump. The water pump can be installed on model C tractors which were originally equipped with a thermo-siphon system.

A. 3 21/32 inches
B. 0.031-0.041
2. Snap ring
3. Seal assembly
4. Impeller
5. Gasket
6. Rear body
7. Front body
8. Shaft and bearing assembly

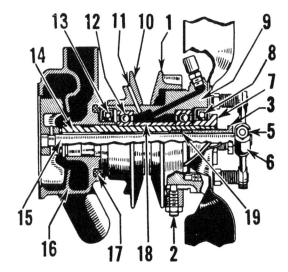

Fig. IH400—Sectional view of the water pump assembly which is used on series H and 4.

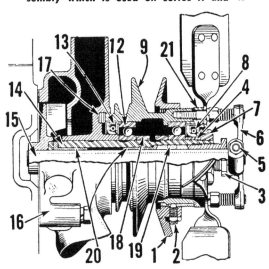

Fig. IH401—Sectional view of the water pump assembly which is used on series M, 6 and 9.

pump shown in Fig. IH400, the adjustable pulley flange (1) is removed in the same manner although it is first necessary to remove the fixed pulley flange pin (11) and unscrew the fixed pulley flange (10) from the hub.

On the H and 4 series, if bushings in bore of sleeve (14—Fig. IH400) are worn it will be necessary to renew the sleeve assembly, as bushings are not furnished separately. On all other models these sleeve bushings (20—Fig. IH401) are furnished separately if desired. The impeller, thrust washer and shaft (15) are furnished only as a single unit. Impeller shaft diameter in Fig. IH400 should measure 0.4355-0.436 and in Fig. IH401 should measure 0.06215-0.622. Clearance of shaft in bushings should be 0.0015-0.0025.

To reassemble pump, press sleeve (14) into pump body (16), making sure that the ⅛-inch hole in sleeve regis-

ters with a similar hole in body. Install rear bearing (12) and oil seal (13) in fan hub (9) with lip of seal facing bearing (adjustable flange (1) and fixed flange (10—Fig. IH400 only) must first be assembled to hub). Insert felt (17) and press the hub assembly on the sleeve and body unit while supporting on end of sleeve. Install spacer (18), front bearing, oil seal (8) with lip facing fan, lock sleeve (4—Fig. IH401 only) and bearing retaining nut (7).

Ball bearings are not preloaded by the spacer (18) so that when fully assembled the front bearing need not be in contact with shoulder in hub. If fan pulley does not revolve freely after bearing retaining nut (7) is fully tightened the probable cause is that spacer (18) is too short, causing the bearings to be preloaded. Correct this with a new spacer.

Reinstall impeller, thrust washer and shaft (15) packing (19) and packing nut (3). Install bearing retainer (21—Fig. IH401 only), fan blades, driver (6) and driver pin (5).

WATER HEADER PLATES
Series H-M-4-6-9

177. Non-Diesel engines are equipped with one water header plate and Diesel engines are equipped with two plates. To remove the front plate, on Diesels, it is necessary to R&R injection pump as outlined in paragraph 133. To remove the rear plate, it is necessary to R&R the fuel filters.

When installing the water header plates on crankcase, the opening in the center of the channel should face downward and the flared end of the channel must face toward the front end of crankcase.

IGNITION AND ELECTRICAL SYSTEM

Early production tractors were factory equipped with a flange mounted IH magneto. The J4 magneto is used on the Cub only. The H4 magneto is used on all other models. Late production tractors are factory equipped with an IH battery ignition unit; magneto ignition being available as optional equipment.

Note: Field change-over kits are available for converting magneto ignition to battery ignition.

DISTRIBUTOR

All Models

178. **INSTALLATION AND TIMING.** To install the battery ignition unit, first crank engine until No. 1 piston is coming up on compression stroke, then continue cranking until the ignition timing marks are in register.

On series A, B and C, the flywheel mark "DC/1-4" or dot mark on flywheel should be in register with indicator (41—Fig. IH405) on clutch housing cover, as viewed through hand hole in bottom of the clutch housing.

On the Cub, the notch on the crankshaft pulley (42—Fig. IH406) should be in register with pointer (43) on the left hand side of the crankcase front cover.

On series H, M, 4, 6 and 9, first (DC) notch (only notch on late H & 4 series) on crankshaft pulley (44—Fig. IH407) should be in register with pointer (45) on crankcase front cover.

Turn the distributor drive shaft until rotor arm is in the No. 1 firing position and mount the ignition unit on the engine, making certain that lugs on ignition unit engage slots in the drive coupling.

Note: If the driving lugs on the battery ignition distributor will not engage the coupling drive slots, when rotor is in No. 1 firing position, it will be necessary to remesh the drive gears as follows: Grasp the distributor drive shaft and pull same outward to disengage the gears. Turn drive shaft until lugs will engage the drive slots, then push drive shaft inward to engage the gears.

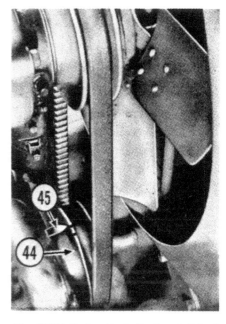

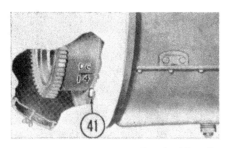

Fig. IH405—Series A, B and C ignition timing marks which can be viewed through hand hole in bottom of clutch housing. Number 1 piston is in top dead center position when flywheel mark "DC/1-4" or dot mark is in register with indicator (41) on clutch housing cover.

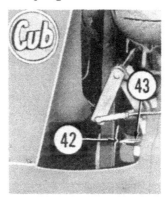

Fig. IH406—Cub ignition timing marks. Number 1 piston is in top dead center position when notch on crankshaft pulley (42) is in register with pointer (43) on left side of crankcase front cover.

Fig. IH407—Series H, 4, M, 6 and 9 ignition timing marks. Number 1 piston is in top dead center position when the first notch on crankshaft pulley (44) is in register with pointer (45) on the crankcase front cover.

179. To time the battery ignition distributor after same is installed as outlined above, proceed as follows: Loosen the distributor mounting cap screws and retard distributor about 30 degrees by turning distributor assembly in same direction as the cam rotates. (Refer to Table 2 for distributor rotation). Disconnect coil secondary cable from distributor cap and hold free end of cable 1/16-1/8 inch from distributor primary terminal. Advance the distributor by turning the distributor body in opposite direction from cam rotation until a spark occurs at the gap. Tighten the distributor mounting cap screws at this point.

To check distributor timing, position the coil secondary cable as outlined above and crank engine slowly. A spark should occur at the gap when timing marks are in register. Assemble the spark plug cables to the distributor cap in the proper firing order of 1-3-4-2.

Running timing can be checked with a neon light, using the advance data given in Table 2.

180. **OVERHAUL.** Defects in the battery ignition system may be ap-

proximately located by simple tests which can be performed in the field; however, complete ignition system analysis and component unit tests require the use of special testing equipment. All of the distributors have automatic spark advance which is obtained by a centrifugal governor built into the unit. Identification and advance curve data are given in Table 2.

MAGNETO
All Models

181. **INSTALLATION AND TIMING.** To install the magneto, first crank engine until No. 1 piston is coming up on compression stroke and ignition timing marks are in register. The timing marks are as follows:

On series A, B and C, the flywheel mark "DC/1-4" should be in register with indicator (41—Fig. IH405) on clutch housing cover, as viewed through hand hole in bottom of the clutch housing.

On the Cub, the notch on the crankshaft pulley (42—Fig. IH406) should be in register with pointer (43) on left hand side of the crankcase front cover.

On series H and 4 and non-Diesel models of the M, 6 and 9 series, the

first notch on crankshaft pulley (44—Fig. IH407) should be in register with pointer (45) on crankcase front cover.

On Diesel models of the M, 6 and 9 series using H4 magnetos with serial numbers prior to 1353301, the notch marked "M" on crankshaft pulley should be in register with pointer on crankcase front cover.

Note: The crankshaft pulleys on some early production M & 6 series tractors were not marked with the letter "M". On these tractors, crank engine until notch marked "DC" is 1½ inch past the pointer.

On Diesel models of the M, 6 and 9 series using H4 magnetos with serial numbers 1353301 and up, the first notch on crankshaft pulley (44—Fig. IH407) should be in register with pointer (45) on crankcase front cover.

Turn the magneto drive lugs until rotor arm is in the No. 1 firing position and mount ignition unit on engine, making certain that lugs on ignition unit engage slots in drive coupling.

182. To time the magneto after same is installed as outlined above, proceed as follows: Loosen the magneto mounting bolts and retard the magneto by turning top of magneto in the normal direction of rotation. (Refer to Table 3 for magneto rotation.) Crank engine one complete revolution and align timing marks as before. Advance magneto by turning top of magneto opposite to normal rotation until impulse coupling snaps; then, tighten the magneto mounting bolts. To check magneto timing, crank engine until No. 1 piston is again coming up on compression stroke and impulse coupling snaps. At this time, the aforementioned timing marks should be in register. Assemble the spark plug cables to the distributor cap in the proper firing order of 1-3-4-2.

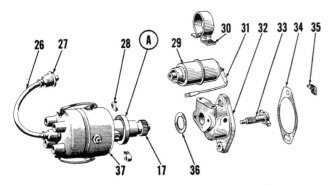

17. Distributor gear
26. Secondary cable
27. Nipple
28. Distributor clamp
29. Coil
30. Coil clamp
31. Primary cable
    (coil to distributor)
32. Drive housing
33. Drive shaft
34. Drive housing gasket
35. Mounting clip
36. Distributor gasket
37. Distributor assembly

Fig. IH409—Exploded view of typical International Harvester battery ignition unit. The distributor identification symbol is stamped on the outside diameter of mounting flange (A).

Fig. IH411—Exploded view of typical International Harvester battery ignition distributor. The distributor identification symbol is stamped on the outside diameter of mounting flange (A). Shaft (13) rides directly in unbushed housing (14)—wear is corrected by renewal of one or both parts.

1. Distributor cap	6. Primary terminal screw	11. Governor weight	17. Distributor gear	21. Weight arm spacer
2. Rotor	7. Insulator	12. Governor spring	18. Thrust washer	22. Cam
3. Breaker cover felt seal	8. Insulating washer	13. Distributor shaft	19. Cap retaining spring	23. Breaker contact set
4. Breaker cover	9. Breaker plate	15. Oil seal retainer	and support	24. Condenser clamp
5. Breaker cover gasket	10. Governor weight guard	16. Oil seal	20. Thrust washer	25. Condenser

Running timing can be checked with a neon light, with engine running faster than the impulse coupling cut-out speed, by using the lag angle data given in Table 3.

183. OVERHAUL. Magneto repairs can be divided into two general classes: Minor repairs and adjustments which can be performed without extensive disassembly of the magneto; and shop overhaul which requires complete magneto disassembly and the use of special tools and testing equipment.

Refer to Table 3 for minor repair data.

IGNITION UNIT MOUNTING BRACKET
Series H-M-4-6-9

184. Bracket and magneto or distributor are mounted on right side of engine, back of crankcase front cover (timing gear cover) and crankcase front plate. The drive shaft rides in a steel-backed, babbitt-lined bushing which must be carefully reamed after installation to provide 0.001-0.003 clearance for the 0.9995-1.0005 diameter drive shaft. Large holes (9/16 inch) in bushing should be vertical and small holes (3/16 inch) should be toward magneto. Front end of bushing should be flush with front face of bracket. Bore of bushing must be square with mounting face of bracket to prevent gear runout. End play of shaft with gear assembled is 0.003-0.013.

When reinstalling the magneto bracket on the engine, time the drive gear to the camshaft gear as per paragraph 66, 67, 67A or 68.

TABLE 3
INTERNATIONAL HARVESTER Magnetos
Rotation is viewed from driving end. Point setting is 0.013 for all magnetos. Breaker arm spring pressure is 23 oz. for the J4 magneto; 19-21 oz. for the H4 magneto.

Tractor Models	Magneto Model	Rotation	Impulse Trips	Lag Angle
A, AV, B, BN, C	H4	C	TDC	35°
Super A, AV, C	H4	C	TDC	35°
Cub	J4	C	TDC	13°
Series H, 4	H4	C	TDC	35°
Series M, 6 (Non-Diesels)	H4	C	TDC	35°
Series 9 (Non-Diesels)	H4	CC	TDC	35°
Series Super M & 6 (Diesels)	H4	CC	TDC	7°
Series 9 (Diesels)	H4*	CC	13° ATC*	15°*
Other Series M & 6 (Diesels)	H4*	CC	6½° ATC*	15°*

*H4 magnetos with serial numbers 1353301 and up have a lag angle of 7 degrees and the magneto must be installed so that the impulse coupling trips at TDC.

TABLE 2
INTERNATIONAL HARVESTER Ignition Distributors
Rotation is viewed from driving end. Advance data are given in distributor degrees and distributor rpm. For flywheel degrees and rpm, double the listed figures. Point setting is 0.020 for all distributors. Breaker arm spring pressure is 21-25 ounces for all distributors.

Tractor Models	Complete Ignition Unit Part Number	Distributor Part Number	Distributor Symbol *	Rotation	No. Cyl.	Start Advance Degrees @ rpm	Intermediate Advance		Full Advance Degrees @ rpm
							Degrees @ rpm	Degrees @ rpm	
A, AV, B, BN, C Super A & AV prior FAAM 349320	353 870 R91	353 890 R91	A	C	4	0-200	9.5-400	17-600	20-800
Super C, other Sup. A&AV	357 934 R91	357 935 R91	J†	C	4	0-200	4.5-400	12-600	15-800
Cub	353 871 R91	353 893 R91	D	C	4	0-200	2.5-400	6.0-600	8-800
H, HV, M, MV	353 870 R91	353 890 R91	A	C	4	0-200	9.5-400	17-600	20-800
W4, O4, OS4	353 870 R91	353 890 R91	A	C	4	0-200	9.5-400	17-600	20-800
W6, O6, OS6	353 870 R91	353 890 R91	A	C	4	0-200	9.5-400	17-600	20-800
W9, WR9, WR9S	353 872 R91	353 891 R91	B	CC	4	0-200	9.5-400	17-600	20-800
Super M & 6 Non-Diesels	357 934 R91	357 935 R91	J†	C	4	0-200	4.5-400	12-600	15-800
Super H, HV, W4	357 934 R91	357 935 R91	J	C	4	0-200	4.5-400	12-600	15-800
MD, MDV, WD6, ODS6	356 610 R91	356 657 R91	H	CC	4	0-200	2.5-400	**	**
Super M & 6 Diesels	356 610 R91	356 657 R91	H	CC	4	0-200	2.5-400	**	**
Series 9 Diesels	356 610 R91	356 657 R91	H	CC	4	0-200	2.5-400	**	**

*This is a distributor identification symbol which is the first letter of the code stamped on the outside of the distributor mounting flange.
†Orginally designated with symbol "E". When servicing symbol "E" distributors install advance spring service package No. 358 108 R91, which will change symbol "E" distributors to symbol "J". The new advance curve improves tractor performance below 1200 engine rpm.
**Reaches maximum advance of 4 degrees at 462.5 rpm.

TABLE 4
ROCKFORD CLUTCH SPECIFICATIONS

Tractor Models	I-H Complete Assembly Number Less Driven Plate	Clutch Model Number	Cover Assembly Number	(A) Cover Setting Inches	(B) Lever Height Inches	PRESSURE SPRINGS		
						No. Used	Pounds Test	at Height
Super A, Super AV, Super C	351447R91	9RM	165249	27/32	2 3/8	6	175	@ 1 7/16
A, AV, B, BN	353242R91	9RM	R4534	27/32	2 23/64	6	185 - 194	@ 1 13/16
C	351447R91	9RM	165249	27/32	2 3/8	6	175	@ 1 7/16
Cub	351760R91	6 1/2 RM	165246	5/16	2 7/32	6	112 - 122	@ 1 1/8
H prior FBH391358, HV prior FBH-V391445, W4 prior WBH34101	52900D	10RM	R4259-1	27/32	2 5/32	6	165 - 175	@ 1 7/16
Other H & 4 Series	358555R91	....	165327	27/32	2 5/32	9	165 - 175	@ 1 7/16
Early M & 6 Series	52840D	11RM	R4596	1 3/32	2 21/32	9	189	@ 1 13/16
Late M & 6 Series	357299R91	12RM	165310	1 1/32	2 21/32	12	150 - 160	@ 1 13/16
Series 9 after Ser. 1439	353176R91	12RM	165278	1 1/32	2 29/32	*9	175 - 180	@ 1 39/64

*Data given is for outer springs. 9 inner springs are also used; length is 1 39/64 inch under 72-76 lbs.
(A) Position of pressure plate in relation to base of clutch cover.
(B) Release lever height from friction face of pressure plate to release bearing contacting surface on release levers.

CLUTCH (Models Without Torque Amplifier)
★ *For Models With Torque Amplifier, Refer to page 53* ★

188. Rockford applications are listed in Table 4 on page 49. Auburn cover assembly applications are as follows:

Cub 6501-7
Series A & C *9001-8
*Replaced by No. 100007

ADJUST SPRING LOADED CLUTCH

190. Adjustment to compensate for lining wear is accomplished by adjusting the clutch pedal linkage, *not* by adjusting the position of the clutch release levers.

The clutch is properly adjusted when the clutch pedal free travel is the value specified below.

Clutch Pedal Free Travel (Inches).
A, AV, B, BN.....................1½
Super A & AV 1-1¼
Super C 1 7/16
Cub (Internal adjustment) 1 3/16
Cub (External adjustment) 1
Series H, M, 4, 6..................1⅛
Series 9, model C 1⅝

On models A, AV, B, BN, Super A, Super AV and C (some), clutch pedal linkage is adjusted by nuts (19—Fig. IH430) which are accessible through an opening in the lower rear portion of the clutch housing.

On early model Cub tractors, adjustment of the pedal linkage is accomplished in the following manner: Remove clutch housing hand hole cover (A—Fig. IH415) and remove pin (B). CAUTION: Before removing pin (B) see that the nut and bolt (D) which fastens the release yoke (E)

together is tight, otherwise pin (B) cannot be reinstalled easily. With pin (B) removed, pry the release yoke (E) forward and allow clutch pedal rod (F) to drop to bottom of clutch housing; pry clutch pedal rod to one side so that release yoke can be pushed back out of the way. Unscrew the nut (C) (on some models this nut (C) is not installed as it usually becomes so tight that it cannot be loosened by hand and requires splitting the tractor to loosen it with a wrench) turn clevis on the pedal rod in until correct pedal free travel is obtained and reinstall the pin (B).

On late model Cub tractors, adjustment of the pedal linkage is accomplished by loosening the cap screw (A —Fig. IH416) and rotating the slotted lever (B).

On all other models, the rod is located externally and the adjustment procedure is evident.

On the A, AV, B, BN and Super A & AV and Cub, pedal free travel is measured at the top of pedal. On all other models, free travel is measured from pedal stop horizontally along path of pedal travel.

ADJUST OVER CENTER CLUTCH

193. When pushing forward on the hand lever to engage the clutch, there should be a definite feel of over-center action. A slight pressure should be felt in the hand lever, then a definite release of pressure as the clutch goes into engagement.

To adjust the clutch, remove the hand hole cover from the clutch hous-

ing and with clutch disengaged, turn engine until adjusting ring lock (A— Fig. IH418) is accessible. Release the lock and turn the adjusting ring (B) clockwise one notch at a time, testing with the hand lever each time until the action described above is felt. Re-engage the adjusting ring lock in the next adjacent notch. CAUTION: Do not make adjustment so tight that it requires undue effort to engage the clutch.

REMOVE AND REINSTALL CLUTCH
Series A-B-C-Cub

195. To remove the clutch, it is first necessary to split the torque tube (clutch housing) from engines as follows—

196. **TRACTOR SPLIT.** To split the torque tube (clutch housing) from engine, remove hood and disconnect the steering shaft universal joint. On models equipped with hydraulic "Touch Control", drain the hydraulic cylinder and remove the hydraulic lines unit. Remove oil cup from air cleaner and on all models except the Cub, remove the starting motor. Disconnect fuel lines, heat indicator send-

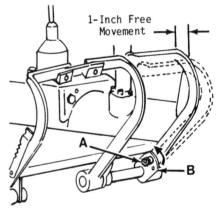

Fig. IH416—Late model Cub clutch pedal linkage adjustment. Adjustment is made by loosening nut (A) and turning lever (B).

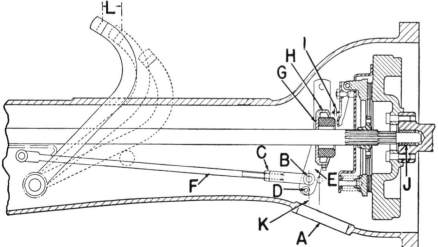

Fig. IH415—Sectional view of Cub clutch housing, showing early model clutch operating linkage.

A. Hand hole cover
B. Yoke pin
C. Clevis locknut
D. Yoke bolt
E. Release yoke
F. Clutch pedal rod
G. Release bearing
H. Bleeder hole
I. Lever to bearing clearance
J. Clutch shaft pilot bushing
K. Yoke travel
L. Pedal free travel

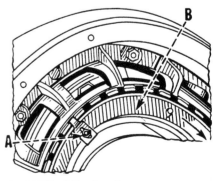

Fig. IH418—Adjusting the over-center type clutch. Adjustment is made by turning adjusting ring (B) after releasing lock (A).

ing unit, wiring harness and controls from engine and engine accessories. Remove the fuel tank front support bolts, loosen the fuel tank rear support and block up between the fuel tank and torque tube or hydraulic cylinder. Support both halves of tractor and remove the clutch housing cover. Remove engine to clutch housing bolts and separate the tractor halves.

After splitting the tractor as outlined in the preceding paragraph, remove the clutch cover retaining cap screws and remove the clutch from the flywheel. On some models, alternate clutch cover retaining cap screws can be used to unload the cover assembly prior to removal.

Series H-M (Except Super MTA)

197. *Late production M, MV, MD and MDV and all Super M, MD, MV and MDV tractors are equipped with a 12-inch diameter clutch. Early production M, MV, MD and MDV tractors were equipped with an 11-inch clutch. The early production 11-inch clutch can be removed as outlined in paragraph 198. The late production 12-inch clutch can be removed as outlined in paragraphs 199 & 200. The removal procedure is not affected by the clutch size on H series tractors.*

198. *To remove the 11-inch clutch which is used on early production series M tractors and either the 10-inch or 10½-inch clutch as used on series H tractors, proceed as follows:*

Remove the cover plate or the hydraulic power lift from the bottom of the clutch housing as per paragraph 520. Remove the clutch release fork

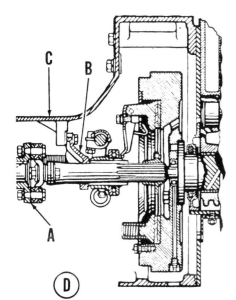

Fig. IH420—Sectional view of series H and M clutch housing and clutch.
A. Clutch universal joint   D. Opening for clutch
B. Release sleeve carrier   removal
C. Clutch housing

shaft and disconnect the clutch-to-transmission universal joint (A—Fig. IH420) and unbolt the release sleeve carrier (B) from the clutch housing (C).

Remove three alternate clutch-to-flywheel cap screws and screw them into tapped holes provided in clutch cover plate until pressure springs are compressed, then remove the three remaining cap screws and remove the clutch shaft, release sleeve carrier and clutch from the tractor.

When reinstalling, reverse above procedure and use the three cap screws in the tapped holes of cover plate to keep springs compressed until clutch is fastened to flywheel.

199. *To remove the clutch cover assembly which is used with the 12-inch diameter lined clutch disc on late production M, MV, MD, MDV and all Super M, MV, MD and MDV tractors, it is necessary to disconnect the clutch housing from the engine. The lined disc, however, can be removed through the opening in the bottom of the housing.*

To remove only the lined disc proceed as follows: Remove the clutch housing cover plate (or hydraulic lift as outlined in paragraph 520), remove the clutch release fork shaft, unbolt the clutch-to-transmission universal joint and disconnect the release sleeve carrier from clutch housing. Then unbolt the clutch cover assembly from the flywheel and remove the clutch shaft. Tilt cover assembly to one side and remove the lined disc.

200. To remove the clutch cover assembly, proceed as follows: Remove hood and disconnect the steering shaft universal joint. The steering worm shaft can be removed for convenience. Disconnect the radiator upper support rod. Drain cooling system and disconnect heat indicator sending unit, fuel lines, oil pressure gage line and wiring harness from engine and engine accessories. On Diesel models, remove the starting control rod and the Diesel fuel supply and return lines. Remove the injection pump air cleaner pipe and carburetor air cleaner pipe. Support rear portion of tractor under clutch housing, and engine half of tractor in a hoist. Remove bolts which retain side rails and engine to clutch housing and move engine half of tractor forward or rear half of tractor rearward. Remove clutch from flywheel.

Install clutch by reversing the removal procedure.

Series 4-6-9 (Except Super W6TA)

204. The clutch assembly is removed

through an opening in top of tractor main frame after removing the clutch compartment cover. To remove the cover, proceed as follows: Remove hood, main fuel tank and engine controls. Remove radiator shutter control by loosening the lock nuts on the steering column, and unscrew shaft by cranking the control handle to left. Remove belt pulley assembly and starting motor, and disconnect oil pressure and heat indicator gage lines. Disconnect brace rods or brackets which interfere, and remove air cleaner and auxiliary fuel tank as a unit; then, remove the cover. Note: On 4 and 6 series tractors, it may be desirable to remove clutch compartment cover, fuel tank support, air cleaner, auxiliary fuel tank and starter as a unit.

On series 4 and 6, remove bolts from clutch shaft coupling, release sleeve carrier and clutch release fork; then, remove the clutch release fork shaft. On spring loaded type clutches, remove three alternate clutch cover-to-flywheel cap screws and screw the screws into the tapped holes in the clutch pressure plate to unload the clutch; then, remove the remaining three cap screws which retain clutch cover to flywheel. On over-center type clutches, remove all of the cap screws retaining the clutch assembly to the flywheel. On all tractors of the 4 and 6 series, rotate clutch shaft ¼ turn and remove the clutch shaft coupling. Pull clutch shaft rearward until free from pilot bearing, and lift clutch assembly from tractor.

On series 9, remove the clutch release fork shaft and the release fork (1—Fig. IH421). Remove bolts (6) from release sleeve carrier (5) and bolts (4) from transmission flange. Lift out clutch coupling spacer (8—Fig. IH422), rubber spacer disc (9), shaft retainers (2) and coupling ring (3). Remove transmission drive flange (11) and pull clutch shaft (15) out of pilot bearing. On spring loaded type clutches, remove three alternate clutch cover-to-flywheel cap screws and screw the screws into the tapped holes in the clutch pressure plate to unload the clutch; then, remove the remaining three cap screws which retain clutch cover to flywheel. On over-center type clutches, remove all of the cap screws retaining the clutch assembly to the flywheel. On all tractors of the 9 series, pull clutch unit away from flywheel and tilt coupling end of shaft downward and to left side under pulley drive gear compartment. With entire assembly resting on bottom of clutch housing, work the clutch shaft out of the lined plate and withdraw clutch assembly from the compartment. The clutch shaft, release sleeve and carrier assembly can now be removed.

Install clutch by reversing the removal procedure, and on series 9, make

Fig. IH421—Removing series 9 clutch assembly.

1. Release fork
2. Half moon retainers
3. Coupling ring
4. Coupling bolts
5. Release sleeve carrier
6. Carrier cap screws
7. Tapped holes in cover plate
8. Clutch coupling spacer
10. Bolt tab locks
16. Pilot bearing lubricator

Fig. IH422—Series 9 clutch to transmission coupling removed.

2. Half moon retainers
3. Coupling ring
4. Coupling bolts
7. Clutch bolt in tapped hole
8. Spacer
9. Rubber spacer disc
10. Bolt tab locks
11. Drive flange
12. Flange retaining washer
13. Flange retaining bolt
14. Clutch to flywheel cap screws
15. Clutch shaft

certain that lubricator fitting (16—Fig. IH421) is located centrally between any two of the bolts (4) otherwise lubrication of the clutch shaft pilot bearing cannot be accomplished.

OVERHAUL SPRING LOADED CLUTCH

206. Overhaul specifications for Rockford clutches are given in Table 4 on page 49. On Auburn clutches proceed as follows:

COVER ASSY. 6501-7. Pressure springs should test 128-142 lbs. at a height of 1¼ inches. To adjust release levers, bolt cover assembly to flywheel with new lined plate in position. Distance from friction face of flywheel to release bearing contacting face of release levers is $2\frac{3}{16}$ inches.

COVER ASSY. 9001-8 or 100007. Pressure springs should test 176-194 lbs. at a height of $1\frac{23}{32}$ inches. To adjust release levers, bolt cover assembly to flywheel with new lined plate in position. Distance from friction face of flywheel to release bearing contacting face of release levers is 2 45/64 inches.

OVERHAUL OVER-CENTER CLUTCH

207. The over-center cam type clutches can be disassembled as shown in Fig. IH424. Clean and check all parts for wear; renew bushings (21) in the sleeve (12) if worn; check the cams (19), cam blocks (17) in pressure plate (14) for wear; check condition of return springs (20); check friction surface of pressure plate (14) and renew if warped, grooved or heat checked; check splines in hub of friction disc (15) and the friction facings; check threads on adjusting ring (4) and back plate (13) for burrs and clean threads if necessary; check links (18) and pins for wear.

When reassembling, reverse the disassembly procedure and be sure return

springs (20) are correctly installed. Before reinstalling clutch to engine, check flywheel for high spots, ridges or heat checks and reface or renew if necessary. Adjust the clutch as per paragraph 193.

CLUTCH RELEASE BEARING

209. The clutch release bearing and/or carrier (sleeve) on all models can be renewed when clutch is removed or tractor is split.

RENEW CLUTCH SHAFT
Series A-B-C

212. Renewal of the clutch shaft requires disconnecting the torque tube (clutch housing) from the rear frame (transmission and differential housing). On series A and B, the disconnecting procedure is obvious. On series C, proceed as follows: Disconnect both "Touch Control" rods (if tractor is so equipped) and governor control rod at their rear attaching points. Disconnect the tail light wire and the steering shaft universal joint. Remove the battery and battery box. Remove the instrument panel and steering shaft support attaching cap screws and slide the instrument panel and steering shaft support assembly forward enough to clear the rear frame top cover. Disconnect the clutch rod, support engine half of tractor and install a rolling floor jack under rear frame. Unbolt rear frame from torque tube and roll the rear portion of tractor rearward. On all models, disassemble the universal joint and remove the clutch shaft.

Cub

213. The clutch shaft is integral with the transmission spline shaft. Refer to paragraph 247.

All Except Series A-B-C-Cub

214. The clutch shaft is removed with the clutch lined plate. Refer to paragraph 197, 198, 199, or 204.

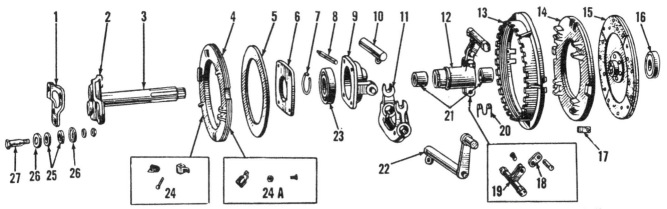

Fig. IH424—Exploded view of series 4 over-center type clutch. Series 6 and 9 over-center clutches are similar.

1. Clutch joint ring
2. Clutch brake facing
3. Clutch shaft
4. Adjusting ring
5. Adjusting ring plate
6. Brake plate
7. Snap ring
8. Bearing lubricator pipe
9. Bearing carrier
10. Release shaft L.H.
11. Release fork
12. Release sleeve
13. Back plate
14. Pressure plate
15. Driven plate
16. Pilot bearing
17. Cam block
18. Connecting link
19. Camshaft
20. Return spring
21. Release sleeve bushings
22. Release shaft R.H.
23. Release bearing
24. Adjusting ring lock (tractors 501 to 1796)
24A. Adjusting ring lock (tractors 1797 up)
25. Clutch joint washer
26. Rubber washer retainer
27. Clutch joint screw

ENGINE CLUTCH (Models With Torque Amplifier)

★ For Models Without Torque Amplifier, Refer to page 50 ★

All models are equipped with spring loaded type clutches manufactured by either Rockford or International-Harvester. The two clutch cover assemblies are interchangeable and the adjustment and overhaul procedures are the same.

ADJUSTMENT

Series MTA-W6TA

214A. Adjustment to compensate for lining wear is accomplished by adjusting the clutch pedal linkage, NOT by adjusting the position of the clutch release levers.

The engine clutch linkage and the torque amplifier clutch linkage should be adjusted at the same time. The adjustment procedure is given in paragraph 215.

REMOVE AND REINSTALL

Series MTA-W6TA

214B. To remove the clutch, it is first necessary to split (detach) engine from clutch housing as outlined in the following paragraph. The clutch can then be unbolted and removed from flywheel in the conventional manner.

214C. **TRACTOR SPLIT.** To detach engine from clutch housing, first remove hood and on series W6TA, disconnect the steering drag link. On series MTA, disconnect the steering shaft universal joint and remove the U-joint Woodruff key. On all models, remove the radiator support rod. Disconnect the fuel lines, wiring harness and controls from engine and engine accessories. On Diesel models, remove the starting control rod and the Diesel fuel supply and return lines. Remove the air cleaner pipes.

Drain the hydraulic system and disconnect the hydraulic lines connecting the pump to the reservoir.

Support tractor under clutch housing and on models so equipped, disconnect the front axle stay rod.

Attach hoist to engine in a suitable manner, remove bolts retaining side rails and engine to clutch housing and move engine half of tractor forward or rear half of tractor rearward.

OVERHAUL CLUTCH

Series MTA-W6TA

214D. Overhaul specifications are as follows: Refer to Fig. IH424A.

Pressure Springs

Test load (lbs.) 130-144

Test length (Inches) ...... $1\frac{13}{16}$

The release lever height (B) from release bearing contacting surface of release levers to friction surface of pressure plate should be $2\frac{5}{16}$ inches when distance (A) from underside of cover plate to friction surface of pressure plate is 1 1/64 inches.

CLUTCH RELEASE BEARING

Series MTA-W6TA

214E. The procedure for renewing the clutch release bearing is evident after detaching engine from clutch housing as outlined in paragraph 214C.

CLUTCH SHAFT

Series MTA-W6TA

214F. To remove the clutch shaft, first split tractor as outlined in paragraph 214C and remove the clutch release bearing and shaft. Unbolt bearing cage (26-Fig. IH424B) from clutch housing and withdraw the independent power take-off drive shaft (25) and clutch shaft (24). Note: A tapped hole is provided in the end of the clutch shaft to aid in its removal.

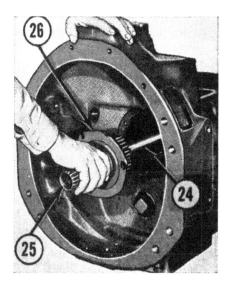

Fig. IH424B—Removing bearing cage (26), independent power take-off drive shaft (25) and clutch shaft (24) from front face of clutch housing.

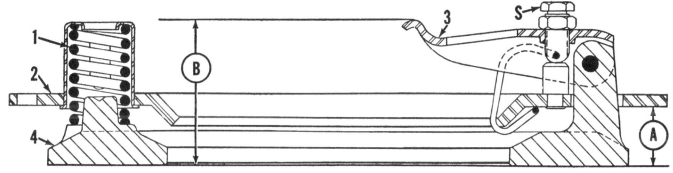

Fig. IH424A—Sectional view of a Rockford spring loaded type clutch showing the release lever adjustment dimension.
A. Position of pressure plate in relation to cover plate which must be maintained when adjusting the release lever height (B).

S. Adjusting screw   1. Pressure spring   2. Cover plate   3. Release lever   4. Pressure plate

TORQUE AMPLIFIER UNIT

Fig. IH425—The International Harvester torque amplifier unit is located between the engine clutch and the transmission.

Torque amplification is provided by a planetary gear reduction unit located between the engine clutch and the transmission as shown in Fig. IH425. The unit is controlled by a hand operated, single plate, spring loaded clutch. When the clutch is engaged as in Fig. IH425A, engine power is delivered to both the primary sun gear (PSG) and the planet carrier (PC). This causes the primary sun gear and the planet carrier to rotate as a unit and the system is in direct drive. When the clutch is disengaged as shown in Fig. IH425B, engine power is transmitted through the primary sun gear to the larger portion of the compound planet gears (PG), giving the first gear reduction. The second gear reduction is provided

by the smaller portion of the compound planet gears driving the secondary sun gear (SSG). As a result of the two gear reductions, an overall gear reduction of 1.482:1 is obtained.

T. A. CLUTCH
Series MTA-W6TA

215. **ADJUST.** Before attempting to adjust the torque amplifier clutch, first make certain that the engine clutch pedal linkage is adjusted properly, as follows:

Refer to Fig. IH426. On series W6TA tractors, remove battery and battery box. On all models, remove spring (6), loosen lock nut (1) and remove clevis pin (3). Loosen lock nut (9) and remove clevis pin (11). Turn clevis (10) until the clutch pedal free travel (N) is ⅞ inch for series

MTA, $\frac{11}{16}$ inch for series W6TA. Dimension (N) is measured horizontally from point of contact of clutch pedal lever and rear frame cover. After adjustment is complete, tighten lock nut (9). Loosen lock nut (13) and turn the adjusting set screw (12) until the pedal full travel (M) is 4 inches for series MTA, $2\frac{13}{16}$ inches for series W6TA. Dimension (M) is measured horizontally from point of contact of clutch pedal lever and rear frame cover. When adjustment is complete, tighten lock nut (13).

After the engine clutch pedal linkage is properly adjusted, place the torque amplifier control lever in the forward position as shown and proceed to adjust the torque amplifier clutch linkage as follows:

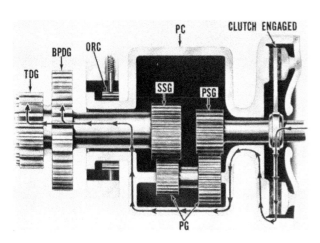
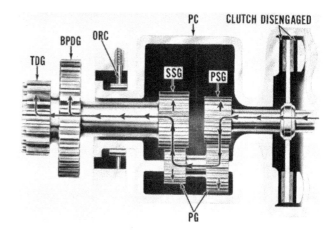

Fig. IH425A—When the torque amplifier clutch is engaged, the system is in direct drive.   Fig. IH425B—When the torque amplifier clutch is disengaged, an overall gear reduction of 1.482:1 is obtained.

Loosen lock nut (5), remove clevis pin (7) and turn lever (4) counter clockwise as far as possible without forcing. See inset. This places the TA clutch release bearing against the clutch release levers. Now, adjust clevis (8) to provide a space (P) of $\frac{3}{16}$ inch between the inserted pin (7) and the forward end of the elongated hole in clevis (8). Tighten locknut (5) and reinstall spring (6). Adjust the length of rod (14) with clevis (2) so that rod (14) is the shortest possible length that will not change the position of levers (4 & 15) when pin (3) is inserted.

216. R & R AND OVERHAUL. To remove the torque amplifier clutch cover assembly and lined plate, first detach (split) engine from clutch housing as outlined in paragraph 214C and proceed as follows: Remove belt pulley unit and the clutch housing top cover (16—Fig. IH426A). Disconnect linkage from the engine clutch release shaft (15), loosen the cap screws retaining fork (27) to shaft and bump shaft toward side of housing until Woodruff keys are exposed. Extract Woodruff keys and withdraw the release shaft and the engine clutch release bearing and carrier (28 & 29). Unbolt bearing cage (26) from clutch housing and withdraw the independent power take-off drive shaft (25) and clutch shaft (24). Note: A tapped hole is provided in the end of the clutch shaft to aid in removal. Disconnect linkage from the TA clutch release shaft and remove snap ring (34) from right end of shaft. Withdraw shaft (4), fork (18) and release bearing and carrier (19 & 20). Unbolt the TA clutch cover assembly (35—Fig. IH426B) from carrier and withdraw the clutch cover assembly and lined plate.

Note: If carrier (47—Fig. IH427B) is damaged, bend tang of locking washer (46) out of notch in nut (45), remove nut as shown in Fig. IH427 and bump carrier from splines of the primary sun gear.

Examine the driven plate for being warped, loose or worn linings, worn hub splines and/or loose hub rivets. Disassemble the clutch cover assembly and examine all parts for being excessively worn. The six pressure springs should have a free length of 2 inches and should require 161 lbs. to compress them to a height of 1¼ inches. Renew pressure plate if it is grooved or cracked. Renew back plate (44—Fig. IH426B) if it is worn around the drive lug windows.

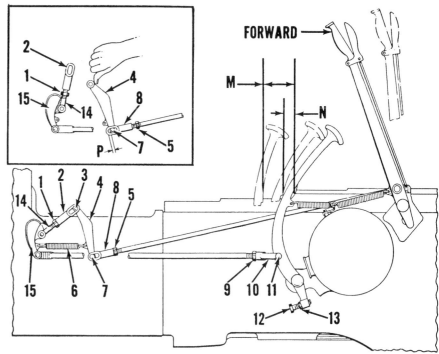

Fig. IH426—Adjusting points for the engine and torque amplifier clutch linkage. Refer to text.

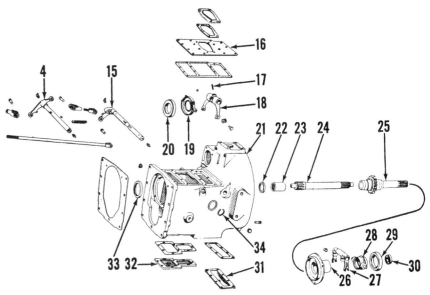

Fig. IH426A—Clutch housing and associated parts used on the Super MTA and Super W6TA tractors. Item (32) is the seasonal disconnect cover for models with independent power take-off.

4. TA clutch release shaft	20. TA clutch release bearing	27. Fork
15. Engine clutch release shaft	21. Housing	28. Engine clutch release sleeve
16. Top cover	22. Oil seal	29. Release bearing
17. Lock screw	23. Coupling	30. Pilot bearing
18. TA clutch release bearing fork	24. Clutch shaft	31. Pto drive gear cover
19. Bearing sleeve	25. Independent pto drive gear and shaft	33. Oil seal
	26. Bearing cage	34. Snap ring

Fig. IH426B—Exploded view of the torque amplifier clutch.

36. Spring cup
38. Pressure plate
39. Driven disc
40. Lever spring
41. Release lever
42. Adjusting screw
43. Lever pin

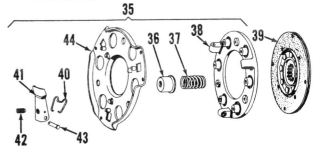

When reassembling, adjust the release levers to the following specifications. With a back plate to pressure plate measurement (K—Fig. IH427A) of $\frac{19}{32}$ inch, the release lever height (L) from friction face of pressure plate to release bearing contacting surface of release levers is $1\frac{5}{8}$ inches.

When reassembling, observe the clutch carrier and back plate for balance marks which are indicated by an arrow and white paint. If the balance marks are found on both parts, they should be assembled with the marks as close together as possible. If no marks are found, or if only one part is marked, the clutch balance can be disregarded. Install the remaining parts by reversing the removal procedure.

Fig. IH427—Using a special OTC socket to remove the torque amplifier clutch carrier retaining nut.

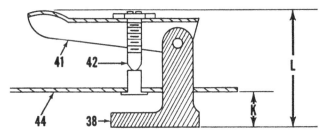

PLANET GEARS, SUN GEARS & OVER-RUNNING CLUTCH
Series MTA-W6TA

217. **R&R AND OVERHAUL.** To overhaul the torque amplifier gear set and over-running clutch, first remove the TA clutch cover assembly, lined plate and clutch carrier as outlined in paragraph 216 and proceed as follows: Remove fuel tank, fuel tank support and air cleaner as an assembly. Disconnect wires and remove starting motor. Remove the independent power take-off seasonal disconnect cover (32—Fig. 426A). Support rear half of tractor under rear frame and attach a chain hoist around clutch housing. Remove bolts retaining clutch housing to transmission case and separate the units. Note: One bolt connecting clutch housing to main frame is accessible through the seasonal disconnect opening.

Unbolt the transmission drive shaft bearing cage from clutch housing and withdraw the complete TA unit. Remove the small retainer ring (70—Fig. IH427B & IH427C) from each of the four Allen head over-running clutch cap screws (71). Clamp the complete unit in a soft jawed vise and remove the four cap screws as shown in Fig. IH427D. Note: A cutout is provided in the planet carrier for this purpose. Separate the transmission drive shaft and bearing cage assembly from planet carrier (78—Fig. IH427B). Remove snap ring (63) from front of transmission drive shaft and press the

Fig. IH427A — When adjusting the release height (L) of 1 5/8 inches, the back plate to pressure plate measurement (K) of 19/32 inch must be maintained.

transmission drive shaft rearward out of bearing and cage. Bearing (64) can be inspected and/or renewed at this time. Inspect ramp (59) springs (68) and rollers of over-running clutch. Renew damaged parts. Using OTC bearing puller attachment 952-A or equivalent, press front and rear bearings (53 & 58) from the planet carrier. It is important, when removing the front bearing to use a piece of pipe and press against the planet carrier and **not** against the primary sun gear

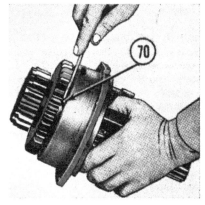

Fig. IH427C—Removing retainer ring (70) from the Allen head over-running clutch cap screws.

Fig. IH427D — Removing the Allen head over-running clutch cap screws.

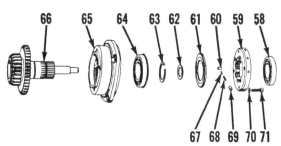

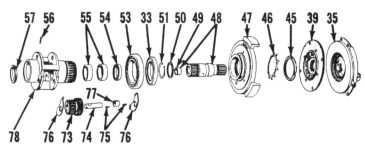

Fig. IH427B—Exploded view of the Super MTA and Super W6TA torque amplifier. Planetary gears (73) are available in sets only.

33. Oil seal	50. Thrust washer	65. Transmission drive shaft
35. Clutch cover assembly	51. Snap ring	bearing cage
39. Driven plate	53. Bearing	66. Transmission drive shaft
45. Nut	54. Oil seal	& secondary sun gear
46. Locking washer	55. Needle bearing	67. Pin
47. Clutch carrier	56. Roll pin	68. Spring
48. Primary sun gear	57. Spacer	69. Plug
49. Roller bearing	58. Bearing	70. Retainer ring
	59. Over-running clutch	71. Allen head screw
	ramp	73. Planetary gears
	60. Clutch roller	74. Gear shaft
	61. Thrust washer	75. Needle bearing
	62. Thrust washer	76. Thrust plate
	63. Snap ring	77. Bearing spacer
	64. Bearing	78. Planet carrier

shaft. Refer to Fig. IH428. Using a small punch and hammer, drive out the Esna roll pins retaining the planetary gear shafts in the planet carrier. Refer to Fig. IH428A. Using OTC dummy shaft No. ED—3259, push out the planet gear shafts and lift gears with rollers and dummy shaft out of the planet carrier. Be careful not to lose or damage the thrust plates (76—Fig. IH427B) as they are withdrawn. After the three compound planetary gears are removed, the primary sun gear and shaft can be withdrawn from the planet carrier.

Inspect splines, oil seal surface, bearing areas, pilot bearing and sun gear teeth of the primary sun gear and shaft for excessive wear or damage. If only the pilot bearing (49) is damaged, renew the bearing. If any other damage is found, renew the complete unit which includes an in-

stalled pilot bearing. Note: When installing a new pilot bearing, use OTC tool No. ED—3251 (shown in Fig. IH428B) and press the bearing in until surface (X) is even with rear edge of primary sun gear.

Inspect the planet gear carrier for rough oil seal surface, worn overrunning clutch roller surface and elongated planet gear shaft holes.

Inspect the primary sun gear shaft needle bearings (55—Fig. IH427B) and the shaft oil seal (54). If bearings and/or seal are damaged, and planet gear carrier is O. K., drift out the faulty parts with a brass drift and install new bearings using OTC driving collar ED—3250 (shown in Fig. IH428C). Press rear bearing in from front until surface (S) is even with front of planet carrier. Press the front needle bearing in from front until surface (T) is even with front of planet carrier. Install oil seal (54—Fig. IH427B) with lip toward rear until front of oil seal is even with front of planet carrier.

Install snap ring (51) on the primary sun gear shaft and thrust washer (50) immediately ahead of the snap ring. Using OTC oil seal protector sleeve No. ED—3245, install the primary sun gear and shaft in the planet carrier as shown in Fig. IH428D.

Inspect teeth of planet gears for wear or other damage. If any one of the three gears are damaged, renew all three gears which are available in a matched set only. These planetary gears are manufactured in matched sets so the gears will have an equal amount of backlash when installed and no one gear will carry more than its share of the load. Note: The International Harvester Co. specifies that

when the planet gears are removed, the planet gear shafts and needle rollers should always be renewed. The planet gear shafts are available in sets of three and the needle rollers are available in sets of 138.

Using chassis lubricant and OTC dummy shaft ED — 3259, install twenty-three new needle bearings in one end of a planet gear. Slide dummy shaft into bearings and install bearing spacer (77—Fig. 427B). With the aid of chassis lubricant, install twenty-three new needle bearings in the other end of the planet gear and slide the dummy shaft completely into the gear, thereby holding the needle bearings and spacer in the proper position. Assemble one thrust plate (76) to each end of the planet gear and install planet gear, dummy shaft and thrust plates assembly in the planet carrier. Using one of the three new planet gear shafts, push out the dummy shaft and install the Esna roll pin securing planet gear shaft to the planet carrier. Refer to Fig. IH428E. Observe the rear face of the planet carrier at each planet gear location where punched

Fig. IH428—Using a piece of pipe and a press to remove front bearing from the planet carrier.

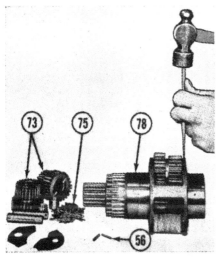

Fig. IH428A—Using a punch and hammer to drive out the Esna roll pins which retain the planet gear shafts in the planet carrier.

56. Roll pins   75. Needle bearings
73. Planet gears   78. Planet carrier

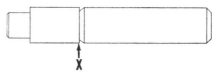

Fig. IH428B—OTC tool number ED-3251. Press pilot bearing in the primary sun gear until surface (X) is even with rear edge of the primary sun gear.

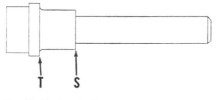

Fig. IH428C—OTC tool number ED-3250 is used to install needle bearings in the planet carrier. Refer to text.

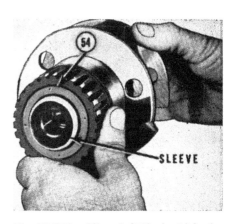

Fig. IH428D—Oil seal (54) should be installed with lip toward rear and front of seal should be even with front of planet carrier. The seal protector sleeve is used when installing the primary sun gear in the planet carrier.

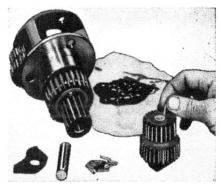

Fig. IH428E—Chassis lubricant facilitates installation of needle bearings in planetary gears.

timing marks will be found. One location has one punch mark, another location has two punch marks and the other location has three. Turn the primary sun gear shaft until the timing punch mark (or marks) on the rear face of the installed planet gear are in register with the same mark (or marks) on the planet carrier. Now assemble the other planet gears, needle bearings, spacers, thrust plates and dummy shaft and install them so that timing marks are in register. When all three planet gears are in-

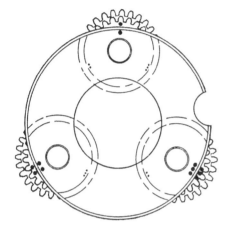

Fig. IH429—When planetary gears are installed properly, punch marks on gears and planetary carrier will be in register.

stalled properly, the single punch mark on planet carrier will be in register with single punch mark on one of the planet gears, double punch mark on carrier will register with double punch marks on one of the gears and triple punch marks on carrier will be in register with triple punch marks on one of the gears as shown in Fig. IH429.

Assemble the pins (67—Fig. IH 429B), springs (68) and rubber plugs (69) into the over-running clutch ramp and place ramp on the planet carrier. Using a small screw driver, push pins (67) back and drop rollers (60) in place as shown. Install bearing (64—Fig. IH429A) with snap ring in the rear bearing cage (65), press the transmission drive shaft (66) into position and install snap ring (63).

Place the planet carrier on the bench with rear end up and lay thrust washer (62) on the primary sun gear. Place the over-running clutch thrust washer (61) on the ramp so that polished surface of thrust washer will contact rollers. Install the assembled transmission drive shaft and bearing cage and secure in position with the four Allen head cap screws (71). After the cap screws are tightened to a

torque of 40 Ft.-Lbs., install the small retainer rings (70).

Inspect the large oil seal (33) in the clutch housing and renew if damaged. Lip of seal goes toward rear of tractor.

Using OTC oil seal protector sleeve No. ED—3253 over splines of planet carrier, insert the assembled TA unit and tighten the transmission drive shaft bearing cage cap screws securely. Assemble the remaining parts by reversing the disassembly procedure.

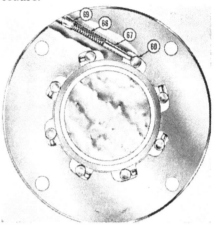

Fig. IH429B—Cut-away view shows proper installation of over-running clutch rollers (60), pins (67), springs (68) and rubber (69) in ramp.

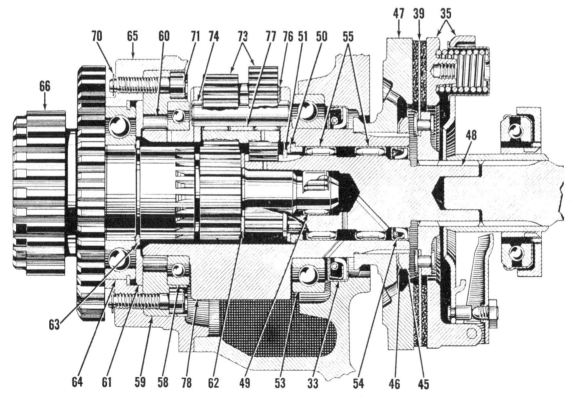

Fig. IH429A—Sectional view showing proper installation of torque amplifier and components.

33. Oil seal	49. Roller bearing	58. Bearing	70. Retainer ring
35. Clutch cover assembly	50. Thrust washer	59. Over-running clutch	71. Allen head screw
39. Driven plate	51. Snap ring	ramp	73. Planetary gear
45. Nut	53. Bearing	60. Clutch roller	74. Gear shaft
46. Locking washer	54. Oil seal	61. Thrust washer	76. Thrust plate
47. Clutch carrier	55. Needle bearing	62. Thrust washer	77. Bearing spacer
48. Primary sun gear		63. Snap ring	78. Planet carrier
		64. Bearing	
		65. Transmission drive shaft	
		bearing cage	
		66. Transmission drive shaft	
		& secondary sun gear	

TRANSMISSION

Although most transmission repair jobs require overhauling the complete unit, there are infrequent instances where the failed or worn part is so located that repair can be accomplished without complete disassembly of the transmission. When such cases occur, considerable time will be saved by following the procedure given under the heading of "Basic Procedures" for the respective tractors.

Series A-B

220. *The transmission and differential assemblies are both contained in the transmission case to which are attached final drive and differential shaft housing assemblies. A wall in the case separates the differential from the transmission gear set. Shifter rails and forks are assembled as a unit attached to the transmission case cover. All bearings in the transmission unit are of the non-adjustable type.*

221. **BASIC PROCEDURES.** Preliminary data for removing the various transmission components is given in the following paragraphs.

222. *Shifter Rails and Forks.* The shifter rails and forks can be removed from the transmission cover after removing the cover from tractor.

223. *Spline Shaft.* The transmission spline shaft (13—Fig. IH430) can be removed after removing the cover or power take-off and/or belt pulley assembly from rear of transmission case, removing the transmission top cover and disconnecting the transmission case from the clutch housing (torque tube).

224. *Countershaft (Bevel Pinion Shaft).* The countershaft and integral main drive bevel pinion (30—Fig. IH430) can be removed after removing the spline shaft as outlined in the preceding paragraph and removing the final drive and differential shaft assemblies and the differential unit.

225. *Reverse Idler.* The reverse idler gear and shaft can be removed after removing the spline shaft as outlined in paragraph 223.

226. **MAJOR OVERHAUL.** Data for removing and overhauling the various transmission components is outlined in the following paragraphs.

227. SHIFTER RAILS AND FORKS. To remove transmission case cover and shifter rails and forks, remove cap screws retaining cover to transmission case. Methods of checking and overhauling shifter rails and forks are conventional. Shifter rails may be removed by spreading the four cap screw locks and removing the cap screws which retain the two shift rail guide brackets to the cover.

228. SPLINE SHAFT. To remove main spline shaft (13—Fig. IH430) proceed as follows: Disconnect transmission case from clutch housing and remove rear cover or power take-off drive housing. Refer to paragraph 499. Remove transmission top cover. Disconnect transmission drive flange (16) and using a suitable puller, remove flange from spline shaft. Remove front bearing retainer (15) and riveted pin which positions reverse gear (7) on spline shaft. Buck up reverse gear and drive main spline shaft (13) rearward and out of case.

When reinstalling, be sure to insert and rivet the pin in reverse gear (7). If special tool for this work is not available, a bar can be bent to hook over side of case to buck up while peening rivet.

229. COUNTERSHAFT. The integral countershaft and bevel pinion (30—Fig. IH430) can be removed after removing the main spline shaft, as in paragraph 228, and differential unit as outlined in paragraph 453.

Remove the front bearing retainer (20) and bearing retaining cap screw (21) and washer. Bump shaft rearward and out of transmission case. Remove bearing cage (22) and shims (A) from front of case and save shims for reinstallation.

When reinstalling countershaft, use original number of shims (A) under bearing cage (22). The fore and aft position of the bevel pinion is controlled by these shims. Check and adjust mesh of bevel pinion and ring gears as outlined under Main Drive Bevel Gears, paragraphs 300 and 301.

230. REVERSE IDLER. To remove the reverse idler gear and shaft, remove main spline shaft as outlined in paragraph 228. Remove the lock screw that retains shaft in position and bump shaft forward. (Models prior to 40429 used a soft plug to retain shaft in position instead of a lock screw.)

Replacement bushings are pre-sized to provide 0 002-0.004 clearance between bushing and reverse idler shaft.

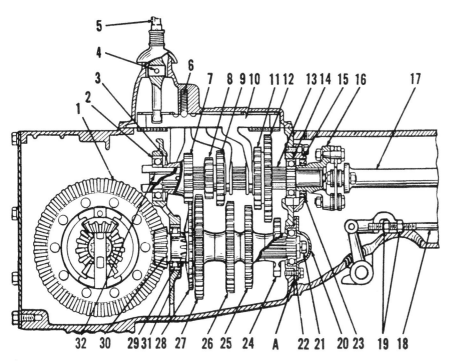

Fig. IH430—Series A and B transmission and differential side sectional view. The countershaft is integral with the main drive bevel pinion.

A. Shims	8. First speed gear	17. Clutch shaft	24. Fourth speed gear
1. Main drive bevel ring gear	9. Second speed gear	18. Clutch actuating rod	25. Third speed gear
2. Bearing	10. Shifter rails	19. Clutch adjustment	26. Second speed gear
3. Oil retainer	11. Third speed gear	20. Countershaft bearing retainer	27. First speed gear
4. Gear shift swivel pin	12. Fourth speed gear	21. Retaining cap screw	28. Oiler gear
5. Gear shift lever	13. Spline shaft	22. Bearing cage	29. Roller bearing
6. Detent	14. Bearing	23. Oil seal	30. Countershaft and bevel pinion
7. Reverse gear	15. Bearing retainer		31. Spacer
	16. Transmission drive flange		32. Bushing

Cub

The transmission gear set and differential unit are both contained in the same case (transmission case). A wall in the case separates the differential unit from the transmission gear set. The differential shaft housing and rear axle housings are bolted to the transmission housing. Shifter rails and forks are mounted in the top of the transmission case. Refer to Fig. IH432. Notice that the transmission spline shaft is integral with the clutch shaft.

240. BASIC PROCEDURES. Preliminary data for removing the various transmission components is given in the following paragraphs.

241. *Shifter Rails and Forks.* The shifter rails and forks can be removed from the transmission case after removing the transmission case rear cover or belt pulley and power take-off assembly, removing the transmission case top cover and disconnecting the transmission case from the clutch housing (torque tube).

242. *Spline Shaft and Integral Clutch Shaft.* The transmission spline shaft (21—Fig. IH432) and integral clutch shaft can be removed after removing the shifter rails and forks as outlined in the preceding paragraph.

243. *Countershaft (Bevel Pinion Shaft).* The countershaft and integral main drive bevel pinion (35—Fig. IH432) can be removed after removing the spline shaft as outlined in the preceding paragraph and removing the final drive and differential shaft assemblies and the differential unit.

244. *Reverse Idler.* The reverse idler gear and shaft can be removed after removing the spline shaft as outlined in paragraph 242.

245. MAJOR OVERHAUL. Data for removing and overhauling the various transmission components are outlined in the following paragraphs.

246. SHIFTER RAILS AND FORKS. To remove the shifter rails and forks, first remove the transmission case rear cover or power take-off, remove the gear shift lever housing (transmission case top cover), detach transmission housing from clutch housing (torque tube) and proceed as follows: Remove cap screws retaining shifter forks to rails and bump shifter rails forward and out of transmission case. As each rail is being removed, insert a short bar into poppet ball hole. This will keep ball in its bore and prevent its flying out or falling into gear case. Expansion plugs will be removed as rails are driven forward.

247. SPLINE SHAFT AND INTEGRAL CLUTCH SHAFT. To remove the transmission spline shaft and integral clutch shaft (21—Fig. IH432), remove the shifter rails and forks as outlined in paragraph 246 and proceed as follows: Remove cap screws retaining spline shaft oil seal retainer (20) to transmission case. Buck up first and reverse speed gear (13) and bump spline shaft forward to remove rear bearing (8) from shaft. Gears will be removed as shaft is withdrawn forward from case.

248. COUNTERSHAFT. The countershaft and integral main drive bevel pinion (35—Fig. IH432) can be removed after removing the spline shaft as outlined in paragraph 247 and the differential as outlined in paragraph 453.

Remove countershaft bearing retainer cap (22) and nut (23) from forward end of shaft. Buck up reverse speed gear (32) and bump shaft rearward. Gears and spacers will be removed as shaft is driven rearward.

Shims (A) located between bevel pinion and countershaft bearing retainer (39) and transmission case are used in setting the main drive bevel pinion gear mesh or cone center distance. Refer to paragraphs 300 and 301 for method of setting the cone center distance and backlash of main drive bevel gears.

249. REVERSE IDLER. To remove the reverse idler gear and shaft, first remove the spline shaft as outlined in paragraph 247. Remove cap screw retaining reverse idler gear shaft in position. Bump shaft forward and out of case, removing the gear as shaft is driven forward. The reverse idler gear bushing can be renewed at this time.

Series C

The transmission, differential and final drive gears are carried in one common case (rear frame). A wall in the rear frame separates the transmission compartment from the differential and final drive compartment. The shifter rails and forks are attached to the rear frame top cover. Bearings are non-adjustable.

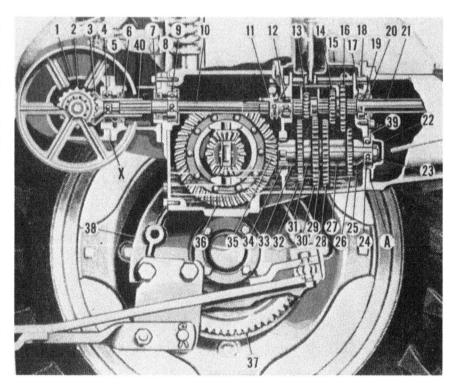

Fig. IH432—Sectional view of Cub transmission and differential assembly. The spline shaft is integral with the clutch shaft.

A. Shims		31. Spacer
X. Shims		32. Reverse speed gear
1. Belt pulley carrier	9. Bearing	19. Snap ring	33. Spacer
2. Belt pulley driven gear	10. Power take-off shaft	20. Oil seal	34. Bearing
		21. Spline shaft	35. Bevel pinion & countershaft
3. Belt pulley drive gear	11. P.T.O. shifter clutch	22. Bearing cap	
4. Bearing	12. Bearing	23. Nut	36. Main drive bevel ring gear
5. Belt pulley drive housing	13. First and reverse speed gear	24. Bearing	
		25. Spacer	37. Rear axle drive gear
6. Oil seal	14. Second speed gear	26. Third speed gear	38. Rear axle housing
7. Oil seal	15. Transmission cover	27. Spacer	39. Bearing retainer
8. Retainer	16. Third speed gear	28. Second speed gear	40. Belt pulley drive gear sleeve
	17. Bearing	29. Spacer	
	18. Shifter rail	30. First speed gear	

252. BASIC PROCEDURES. Preliminary data for removing the various transmission components is given in the following paragraphs.

253. *Shifter Rails and Forks.* The shifter rails and forks can be removed from the rear frame top cover after removing the cover from tractor.

254. *Spline Shaft.* The transmission spline shaft (8—Fig. IH434) can be removed after performing the following work. First remove the rear frame top cover and detach the rear frame from the clutch housing (torque tube). Remove rear frame rear cover or power take-off and/or belt pulley assembly. Loosen the eight nuts on the differential case bolts and bump the main drive bevel ring gear toward left side of tractor far enough to permit spline shaft to pass the teeth on the bevel ring gear.

255. *Countershaft (Bevel Pinion Shaft).* The countershaft and integral main drive bevel pinion (22—Fig. IH434) can be removed after removing the left bull gear, left bull pinion and shaft, left bull pinion shaft bearing cage and the transmission spline shaft.

256. *Reverse Idler.* The reverse idler gear and shaft can be removed after removing the spline shaft as outlined in paragraph 254.

257. MAJOR OVERHAUL. Data on overhauling the various transmission components are given in paragraphs 258, 259, 260 and 261. Before any overhaul work can be accomplished, however, the transmission (rear frame) cover must be removed as outlined in the following paragraph.

257A. Remove the rear frame top cover as follows: Disconnect the "Touch Control" and governor control rods at their rear attaching points. Disconnect the tail light wires and the steering shaft universal joint. Remove battery and battery box. Remove the instrument panel and steering shaft support attaching cap screws and slide the instrument panel and steering shaft support forward enough to clear the rear frame top cover; then, unbolt and remove the cover.

258. SHIFTER RAILS AND FORKS. To remove the shifter rails and forks, first remove the transmission cover as in paragraph 257A. The shifter rails and forks can be removed from the transmission (rear frame) cover by spreading the locks and removing the cap screws which retain the shifter rail guide brackets to the cover. Methods of checking and overhauling the shifter rails and forks are conventional.

259. SPLINE SHAFT. To remove the transmission spline shaft (8—Fig. IH434), remove transmission cover as in paragraph 257A, detach transmission from clutch housing (torque tube) and remove the differential. Remove the cap screw which retains the coupling flange to the spline shaft and using a suitable puller, remove the flange. Remove the Woodruff key from front of spline shaft and remove the spline shaft front bearing retainer (11—Fig. IH434). Remove rivet which positions the reverse gear (6) on the spline shaft, buck up behind the reverse gear, drive the spline shaft out rear of case and remove the gears from above. The bushing in the rear of the spline shaft can be renewed at this time.

Reinstall the spline shaft and gears by reversing the removal procedure and when installing the reverse spline gear rivet, buck up same while peening.

260. COUNTERSHAFT. The countershaft and integral main drive bevel pinion (22—Fig. IH434) can be removed after removing the spline shaft as per preceding paragraph. Remove the countershaft front bearing cage retainer and nut (13). Drive the shaft out rear of transmission case and remove gears from above. Remove bearing cage (15) and shims (A) from front of case and save the shims for reinstallation. When reinstalling the countershaft, use the original number of shims (A) under the bearing cage. The fore and aft position of the bevel

pinion is controlled by these shims. Check and adjust the mesh of the bevel pinion and ring gears as outlined in paragraphs 300 and 301.

260A. REVERSE IDLER. The reverse idler gear and shaft can be removed after removing the transmission spline shaft as in paragraph 259. Remove the reverse idler shaft lock pin and drive the shaft forward. Replacement bushings in the reverse idler gear are presized.

Series H-M-4-6
(Except MTA & W6TA)

For series MTA-W6TA, refer to page 63.

The transmission, differential and bull gears are all contained in the same case which is called the rear frame, as shown in Fig. IH438. A wall in the case separates the bull gears and differential from the transmission gear set. Shifter rails and forks are mounted on underside of the rear frame (transmission) cover. The gear set can be overhauled without splitting the tractor.

All bearings in the transmission unit are of the non-adjustable type.

261. BASIC PROCEDURES. Preliminary data for removing the various transmission components is given in the following paragraphs.

262. *Shifter Rails and Forks.* The shifter rails and forks can be removed from the transmission cover after removing the cover from the tractor.

263. *Main Drive Gear (Pinion) and Shaft (Input Shaft).* The transmission drive shaft and integral pinion gear can be removed after removing the

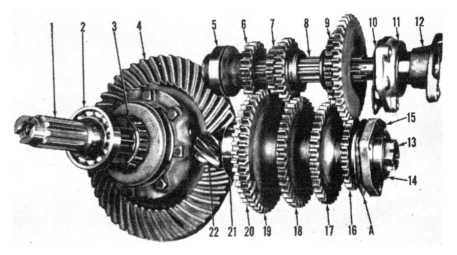

Fig. IH434—Series C transmission shafts and differential assembly. Shims (A) control fore and aft position of the main drive bevel pinion.

1. Bull pinion & brake shaft	7. First & second sliding gear	15. Bearing cage
2. Bearing	8. Spline shaft	16. Fourth speed gear
3. Differential carrier bearing	9. Third & fourth sliding gear	17. Third speed gear
4. Bevel ring gear	11. Bearing retainer	18. Second speed gear
5. Bearing	12. Drive flange	19. First speed gear
6. Reverse spline gear	14. Bearing	20. Oiler gear
		21. Roller bearing
		22. Bevel pinion and countershaft

rear frame (transmission) cover, hydraulic pump unit and clutch shaft.

264. *Main Shaft (Bevel Pinion Shaft)*. The main shaft and integral main drive bevel pinion can be removed after removing the main drive gear and shaft (input shaft) as per preceding paragraph.

265. *Countershaft*. The transmission countershaft can be removed after removing the main shaft as per preceding paragraph and the power take-off unit.

266. *Reverse Idler*. The reverse idler and shaft can be removed after removing the countershaft as per paragraph 265.

268. **MAJOR OVERHAUL**. Data on removing and overhauling the various transmission components are outlined in the following paragraphs.

269. SHIFTER RAILS AND FORKS. First step in disassembly is to remove the rear frame (transmission) cover which is accomplished as follows: Remove belt pulley assembly. On series 4 and 6, remove the steering gear unit

and block up the fuel tank. On series H and M, disconnect the steering post and fuel tank rear support and block up same high enough to clear cover.

Remove seat assembly. Remove cover to rear frame cap screws and lift off cover and shifter mechanism as a unit.

Methods of checking and overhauling shifter rails and forks assembly are conventional.

270. MAIN DRIVE (PINION) GEAR. To remove the main drive gear (15—Fig. IH439) (shaft and two gears are integral), remove rear frame (transmission) cover as in paragraph 269, and the clutch shaft (19) as outlined in paragraph 214. Remove cap screws retaining drive gear bearing cage (16) to case and remove cage and gear assembly. Bearing cage flange has tapped holes into which two of the retaining cap screws can be inserted to act as a puller if cage does not readily come loose from case.

The main shaft pilot bearing (26) can be inspected or renewed at this time. Renew worn parts, reassemble and reinstall in reverse order.

271. MAIN (BEVEL PINION) SHAFT. After removing the transmission main drive (pinion) gear as outlined in preceding paragraph 270, proceed to remove the main drive bevel pinion and shaft (36—Fig. IH-439) (spline shaft) by removing the rear bearing cage (34). Withdraw shaft and gears forward and then out through cover opening. All mainshaft bearings are non-adjustable but the shims (A) located at rear bearing cage should be kept together as these control the mesh position of the bevel drive pinion. The integral main shaft and bevel drive pinion may be purchased separately from the bevel drive ring gear.

Rear bearing (9) can be driven off of shaft with a drift pin inserted through holes provided in the main drive bevel pinion. After reassembling shaft and installing in case, check mesh or cone center distance of bevel pinion and adjust with shims (A), if necessary, as outlined in paragraphs 300 and 301.

272. COUNTERSHAFT. The countershaft (35—Fig. IH439) can be removed as follows after first removing the main (bevel pinion) shaft, paragraph 271: Remove power take-off assembly. Remove front bearing retainer nut (23) and bump shaft rearward to permit removal of rear bearing snap ring (7) which is accessible from differential chamber. Reinstall

Fig. IH438—Series H transmission, differential and final drive assembly. Series M is similar. The Super H and Super M are similar except disc type brakes are used.

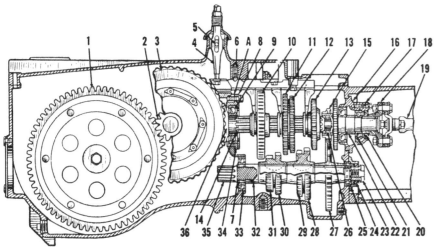

Fig. IH439—Side sectional view of model H transmission. Series HV, other series H and series M, 4 and 6 are similar.

A. Shims		
1. Bull gear	9. Bearing	16. Bearing cage	27. Constant mesh gear
2. Bull pinion	10. First and reverse	17. Bearing retainer	28. Fourth gear
3. Main drive bevel	speed gear	18. Drive shaft coupling	29. Third speed gear
ring gear	11. Second speed gear	19. Clutch shaft	30. Second speed gear
4. Shift lever	12. Third speed gear	20. Oil seal	31. First speed gear
5. Shift lever swivel	13. Fourth and fifth	21. Bearing	32. Spacer
pin	speed gear	22. Spacer	33. Bearing
6. Detent	14. Snap ring	23. Countershaft nut	34. Bearing cage
7. Snap ring	15. Main drive gear	24. Oil seal	35. Countershaft
8. Bearing cage cover	and shaft	25. Bearing retainer	36. Main drive bevel
		26. Pilot bearing	pinion and shaft

nut (23) to pull shaft forward and drive rear bearing (33) off the shaft towards the rear, using an offset drift against the inner race. Remove bearing retainer (25) and nut (23) from forward end of shaft. Bump shaft rearward and out of front bearing. Lift countershaft and gear unit from case. Bump front bearing (25) from case bore and remove the bearing.

Gears are keyed to countershaft and can be pressed on and off shaft. Renew bushing in rear end of shaft if worn or damaged. Bushing bore diameter should be 0.908-0.909 inches.

273. REVERSE IDLER. To remove the reverse idler gear and shaft, first remove the countershaft as per paragraph 272 and proceed as follows: Remove the brake pedal cross shaft. Remove the bolt from the idler gear

shaft bracket, bump the shaft forward and remove the idler gear.

The bushings in the idler gear can be renewed at this time. Ream the bushings if necessary, to provide 0.003-0.005 clearance for the idler gear shaft.

Lock bolt hole through idler shaft is drilled off center to assure correct alignment of grease holes. Check for alignment of hole in shaft and holes in lock bolt boss when reinstalling.

Series MTA-W6TA

The transmission, differential and final drive gears are all contained in the same case which is called the rear

frame. A wall in the case separates the bull gear and differential compartment from the transmission gear set. Shifter rails and forks are mounted on underside of the rear frame (transmission) cover.

All bearings in the transmission are of the non-adjustable ball type.

274. OVERHAUL. Data on removing and overhauling the various transmission components are given in the following paragraphs:

275. REAR FRAME COVER. To remove the rear frame top cover, remove hood, unbolt and block up fuel tank, disconnect the pulley operating rod and remove the belt pulley assembly. Drain the hydraulic system and remove the reservoir and control valves.

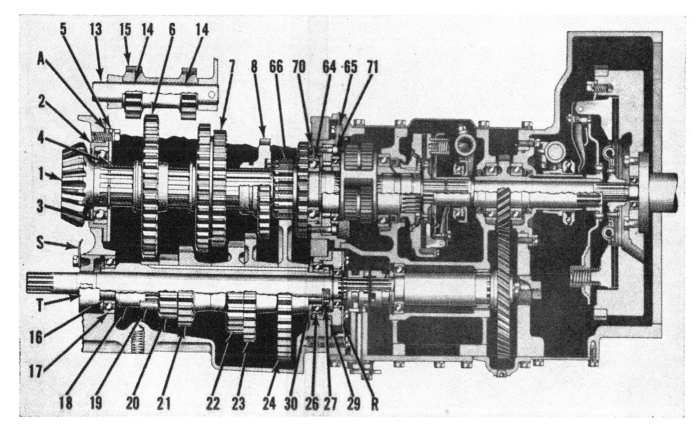

Fig. IH440—Sectional view of Super MTA and Super W6TA transmission. Bearings in the transmission are of the non-adjustable ball type. Fore and aft position of the main drive bevel pinion is controlled by shims (A).

1. Mainshaft & bevel pinion	6. First and reverse sliding gear
2. Bearing cage	7. Second and third sliding gear
3. Rear bearing	8. Fourth and fifth sliding gear
4. Snap ring	
5. Bearing retainer	

13. Reverse idler shaft	18. Countershaft
14. Reverse idler bushing	19. Spacer
15. Reverse idler gear	20. First speed driving gear
16. Snap ring	21. Second speed driving gear
17. Bearing	22. Third speed driving gear
	23. Fourth speed driving gear

24. Constant mesh gear
26. Bearing
27. Nut
29. Bearing cage
30. Spacer

On series W6TA, remove battery cover and disconnect battery cables. Disconnect the steering drag link, wiring harness and control rods and remove steering gear and instrument panel assembly.

On series Super MTA, disconnect battery cables and remove battery and battery box. Disconnect steering shaft universal joint, wiring harness and rods and remove steering wheel and post.

On all models, unbolt cover from rear frame and lift cover from tractor.

276. SHIFTER RAILS AND FORKS. Shifter rails and forks are retained to bottom side of the rear frame cover and are accessible for overhaul after removing the cover. The procedure for disassembling and overhauling is conventional and evident after an examination of the unit.

277. TRANSMISSION D R I V I N G SHAFT. The transmission driving gear and shaft (66—Fig. IH440) is integral with the torque amplifier second sun gear and is normally serviced in conjunction with overhauling the torque amplifier unit as outlined in paragraph 217.

A brief procedure, however for removing the shaft is as follows:

Detach engine from clutch housing and remove the belt pulley unit and clutch housing top cover. Remove the engine clutch release bearing and shaft, unbolt bearing cage from front of clutch housing and withdraw the engine clutch shaft and independent power take-off drive shaft. Remove the torque amplifier clutch release shaft and remove the torque amplifier clutch cover assembly and lined plate. Remove the clutch carrier.

Remove fuel tank, fuel tank support and air cleaner. Remove starting motor and the independent power take-off seasonal disconnect cover. Block-up rear frame, unbolt and remove clutch housing.

Unbolt the transmission drive shaft bearing cage from rear of clutch housing and remove the complete torque amplifier unit. Remove the four Allen head cap screws and separate the transmission drive shaft bearing cage from the planet carrier. The drive shaft can be renewed at this time.

278. DETACH (SPLIT) CLUTCH HOUSING FROM REAR FRAME. Remove hood and fuel tank, drain hydraulic system and remove hydraulic reservoir. Disconnect the battery cables, wiring harness and control rods. Disconnect the steering shaft universal joint or drag link and remove belt pulley assembly. Remove the independent power take-off seasonal disconnect cover from lower side of clutch housing, support rear half of tractor under main frame and attach a chain hoist around clutch housing. Remove bolts retaining clutch housing to main frame and separate the units. Note: One bolt connecting clutch housing to main frame is accessible through the seasonal disconnect opening.

279. MAINSHAFT PILOT BEARING. To remove the mainshaft pilot bearing (9—Fig. IH441), separate the clutch housing from main frame as outlined in paragraph 278. Unlock and remove the two cap screws and washer from front end of mainshaft and using

OTC bearing puller or equivalent as shown in Fig. IH440A, remove the pilot bearing.

280. MAINSHAFT (S L I D I N G) GEAR OR BEVEL PINION SHAFT). To remove the transmission main shaft, remove the rear frame top cover and separate the clutch housing from the rear frame.

Remove the three cap screws retaining the mainshaft rear bearing retainer (5—Fig. IH441) to the main case dividing wall, move the mainshaft assembly forward and withdraw the unit, rear end first, as shown in Fig. IH441A.

Remove the mainshaft pilot bearing (9—Fig. IH441) and slide gears from shaft. Remove snap ring (4) and press or pull rear bearing and retainer from mainshaft. When reassembling, install bearing (3) so that ball loading grooves are toward front or away from the bevel pinion gear. Use Fig.

Fig. IH441A—Removing the transmission mainshaft assembly.

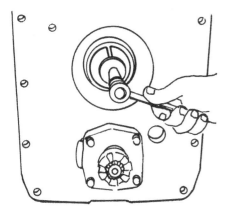

Fig. IH440A—Using OTC puller to remove pilot bearing from front of transmission mainshaft.

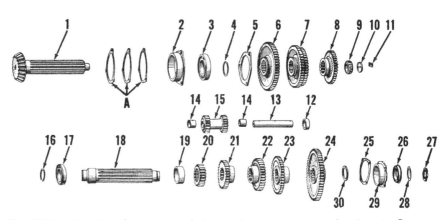

Fig. IH441—Exploded view of transmission shafts, gears and associated parts. Countershaft (18) is hollow on models with independent power take-off. Shaft (18) is splined at rear end on models which have non-continuous power take-off.

A. Shims	8. Fourth and fifth	22. Third speed
1. Mainshaft	sliding gear	driving gear
2. Bearing cage	9. Mainshaft pilot	23. Fourth speed
3. Rear bearing	bearing	driving gear
4. Snap ring	10. Washer	24. Constant mesh gear
5. Bearing retainer	11. Lock plate	26. Bearing
6. First and reverse	13. Reverse idler shaft	27. Nut
sliding gear	14. Reverse idler bush-	28. Locking washer
7. Second and third	ing	29. Bearing cage
sliding gear	15. Reverse idler gear	30. Spacer
	16. Snap ring	
	17. Bearing	
	18. Countershaft	
	19. Spacer	
	20. First speed	
	driving gear	
	21. Second speed	
	driving gear	

IH440 as a guide when installing the sliding gears and if the same mainshaft is installed, be sure to use the same shims (A) as were removed. If a new mainshaft is being installed, use the same shims (A) as a starting point, but be sure to check and adjust if necessary the main drive bevel gear mesh position as outlined in the main drive bevel gear section.

281. COUNTERSHAFT. To remove the transmission countershaft, first remove the mainshaft as outlined in paragraph 280 and proceed as follows: Remove the four cap screws retaining the independent power take-off ex-

Fig. IH441B — Removing the independent power take-off extension shaft, front bearing and cage.

tension shaft front bearing cage to main frame and withdraw the extension shaft, bearing cage and retainer as shown in Fig. IH441B. Working in the bull gear compartment of the main frame, remove the independent power take-off coupling shaft and the extension shaft rear bearing carrier retainer strap (S—Fig. IH440).

Remove nut from forward end of countershaft and while bucking up the countershaft gears, bump the countershaft rearward until free from front bearing. Remove the countershaft front bearing and cage.

Withdraw the countershaft from rear and remove gears from above. The rear bearing can be removed from countershaft after removing snap ring (16—Fig. IH441). When reassembling, use Fig. IH440 as a guide and make certain that beveled edge of constant mesh gear spacer (30) is facing toward front of tractor.

282. REVERSE IDLER. With the countershaft removed as outlined in paragraph 281, the procedure for removing the reverse idler is evident. Bushings (14—Fig. IH441) are renewable and should be reamed after installation, if necessary, to provide a recommended clearance of 0.003-0.005 for the idler gear shaft.

Series 9

The transmission and differential assemblies are both contained in the same case (rear frame). A wall in the case separates the differential from the transmission gear set. Shifter rails and forks are mounted on underside of steering gear housing (rear frame cover). Refer to Fig. IH442.

All bearings in the transmission unit are of the non-adjustable type.

286. **MAJOR OVERHAUL.** The gear set can be overhauled without splitting the tractor (disconnecting transmission from clutch housing) but requires detaching the final drive (rear axle assembly) from the transmission case as shown in Fig. IH464 and described in paragraph 480 to adjust mesh and backlash of main drive bevel gears.

287. SHIFTER RAILS AND FORKS. To remove the shifter rails and forks it is necessary first to remove the steering gear housing (10—Fig. IH-442) (rear frame cover) as outlined in paragraph 21. Methods of checking, overhauling and reassembling the shifter rails and forks are conventional. Fifth speed is locked out in all tractors equipped with steel wheels by a set screw located in the rear frame cover, which engages the shifter rail.

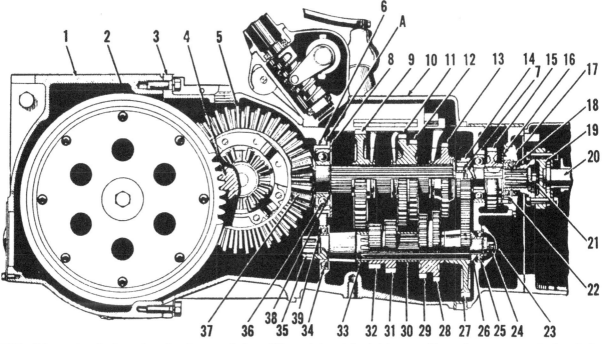

Fig. IH442—Side sectional view of series 9 transmission, differential and final drive. For illustrative purposes the bevel ring gear (5) is shown to the left of the bevel pinion. Actually, the ring gear is installed on the right.

A. Shims	8. Bearing cage cover	20. Clutch shaft	26. Spacer	33. Spacer
1. Final drive housing	9. First and reverse gear	21. Drive flange screw	27. Constant mesh gear	34. Bearing
2. Bull gear	11. Second speed gear	22. Bearing	28. Fourth speed gear	35. Countershaft
3. Transmission case	12. Third speed gear	23. Countershaft front bearing retainer	29. Third speed gear	36. Bearing
4. Bull pinion	13. Fourth and fifth gear	24. Cap screw	30. Spacer	37. Bevel pinion
5. Main drive bevel gear	14. Main drive gear	25. Bearing	31. Second speed gear	38. Snap ring
	15. Pulley drive gear		32. First speed gear	39. Snap ring
	16. Bearing retainer			
	17. Bearing cage			
	18. Oil seal			
	19. Clutch shaft coupling			

288. MAIN DRIVE GEAR (PINION) AND SHAFT (INPUT SHAFT). To remove the main drive gear and shaft (14—Fig. IH442) (gear and shaft are integral) proceed as follows: Remove rear frame (transmission) cover as outlined in paragraph 21. Remove the clutch as in paragraph 204. Remove the cap screws retaining drive gear bearing cage (17) and remove cage and gear assembly. Bearing cage flange has tapped holes into which two of the retaining cap screws can be inserted to act as a puller if cage does not readily come loose from case. When removing bearing cage, insert suitable object in one cap screw hole to prevent cage from turning, as its irregular shape will prevent it from being removed if it has turned. The main shaft pilot bearing (7) can be inspected or renewed at this time.

Renew worn parts, reassemble and reinstall in reverse order with lip of oil seal facing toward gear. Install bearing cage (17) so that the part number on the flange is up.

289. MAIN (BEVEL PINION) SHAFT. After performing work outlined in paragraphs 287 and 288, proceed to remove main drive bevel pinion and shaft (37—Fig. IH442) by removing the rear bearing cage (6).

Withdraw shaft and gears forward and then out through top cover opening. Bearings are non-adjustable but the shims (A) located between rear bearing cage (6) and transmission case wall should be kept together as these control the mesh position of the bevel drive pinion.

Main shaft and bevel drive pinion are integral and may be purchased separately from the bevel ring gear. After reassembling shaft and installing in transmission case, check mesh or cone center distance of bevel pinion and adjust if necessary as outlined in paragraphs 300 and 301 under Main Drive Bevel Gears. Shims (A) are of the half circle type and removal or installation of same can be accomplished without removing the main shaft.

290. COUNTERSHAFT. The countershaft (35—Fig. IH442) can be removed after removing drive pinion shaft, paragraph 288, and main shaft, paragraph 289. Remove the power take-off unit. Remove countershaft front bearing retainer (23—Fig. IH-442) and bearing retaining cap screw (24) and washer from forward end of shaft. Reinstall cap screw in forward end of shaft and while bucking up gear (27) bump shaft rearward until rear bearing (34) is free of case bore

and front bearing (25) is off of shaft. Slide the shaft rearward to remove the gears and spacers. The shaft can be removed by pushing it rearward and removing it through the power take-off opening in the axle housing.

When reinstalling the countershaft use a ⅝ inch—18 cap screw 2 inches long and bearing retaining washer to draw shaft into front bearing and rear bearing into case bore. With the shaft properly installed, replace the 2 inch cap screw with the regular 1¼ inch cap screw (24) and install front bearing retainer (23).

Renew bushing (for power take-off shaft) in rear end of shaft if worn or damaged. Press bushing in until outer end is flush with end of 60 degree chamfer and ream to 1.001-1.002.

291. REVERSE IDLER. To remove the reverse idler shaft and gear, the countershaft must be removed as outlined in paragraph 290. Idler shaft can then be slid forward to free gear after removing bolt from boss extending from side of rear frame.

The two bushings in the idler gear should be pressed in position until flush with chamfer on each end of bore. Bushings require final sizing to 1.616-1.617 to provide 0.003-0.005 clearance.

MAIN DRIVE BEVEL GEARS AND DIFFERENTIAL

ADJUST BEVEL GEARS

All Models

300. The tooth contact (mesh pattern) and backlash of the main drive bevel gears is controlled by shims on all models. Tooth contact (mesh pattern) and backlash should be checked and adjusted, if necessary, whenever the transmission is being overhauled, and always when a new pinion or ring gear, or both, are installed.

301. MESH AND BACKLASH. The first step in adjusting a new set of bevel gears is to arrange shims (B), as shown in the final drive illustrations, to provide the desired backlash between the main drive bevel pinion and ring gear. Desired bevel gear backlash is as follows:

Series A-B 0.005-0.007
Cub 0.003-0.005
C-Super C 0.008-0.010
Series H-M-4-6 .... 0.008-0.012
Series 9 .... 0.010-0.012

The next step is to arrange shims (A), as shown in the transmission illustrations, to provide the proper tooth contact (mesh position) of the bevel gears.

On the Super C tractor, the proper trial setting for the main drive bevel pinion can be obtained by varying the number of shims (A) until a 0.375 rod (X—Fig. IH455) can be inserted between the end of the pinion shaft and the finished diameter of the differential case.

On all models, paint the bevel pinion teeth with Prussian blue or red lead and rotate ring gear in normal direction of rotation under zero (light) load and then observe the contact pattern on the tooth surfaces.

The area of **heaviest contact** will be indicated by the coating being **removed** at such points. On the actual pinion the tooth contact areas shown in black on the illustrations will be bright; that is, there will be no blue or red coating on them.

The desired condition is indicated in Fig. IH445, which shows that the paint has been removed from the toe end of the teeth over the distance A to B as shown.

When the heavy contact is concentrated high on the toe at A as shown in Fig. IH446, the pinion should be

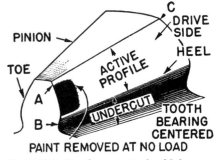

Fig. IH445—Tooth contact should be centered between "A" and "B" (no load). When a heavy load is placed on the gears, the tooth contact will extend from the toe almost to the heel of the tooth.

Fig. IH446—High tooth contact at "A" (no load) indicates that the pinion has been set too far in.

moved toward front of tractor by adding a shim behind the **pinion bearing cage.**

When the heavy contact is concentrated low on the toe of the pinion tooth as in Fig. IH447, remove a shim from the pinion bearing cage.

After obtaining desired tooth contact, recheck the backlash and if not within the desired limits as listed, adjust by shifting a shim or shims from behind one differential bearing cage to the other bearing cage until desired backlash is obtained.

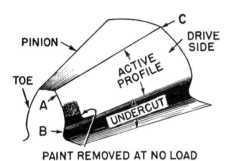

Fig. IH447—Low tooth contact at "B" (no load) indicates that the pinion has been set too far out.

Do not expect the contact pattern to extend much farther toward the heel of the pinion than shown in Fig. IH445. The teeth are purposely ground to produce a toe bearing or contact under zero or light load so that when the gear supports deflect and the teeth deform under heavy load conditions, the contact pattern will **increase in** area along the teeth toward the heel and thus automatically **increase the** load carrying capacity **of the gear.**

RENEW BEVEL GEARS
Series A-B-C-Cub

450. To renew the main drive bevel pinion, follow the procedure outlined for overhaul of the transmission countershaft (paragraph 229 for series A & B), (paragraph 248 for the Cub) and (paragraph 260 for series C). To renew the main drive bevel ring gear, follow the procedure outlined for overhaul of the differential assembly (paragraph 453 for the Cub and Series A & B) and (paragraph 455 for C and Super C).

Adjust the mesh position and backlash of the main drive bevel gears as outlined in paragraphs 300 and 301.

Series H-M-4-6-9

451. To renew the main drive bevel pinion, follow the procedure outlined for overhaul of the transmission main shaft (paragraph 271 for Series H, M, 4 and 6), (paragraph 280 for series MTA and W6TA) and (paragraph 289 for Series 9). To renew the main drive bevel ring gear, follow the procedure outlined for overhaul of the differential assembly (paragraph 458 for Series H, M, 4 and 6) and (paragraph 460 for Series 9).

DIFFERENTIAL AND CARRIER BEARINGS

Cub-Series A-B

Differential is of the two pinion open case type mounted back of a dividing wall in the transmission case. The bevel ring gear is held to the one piece case by rivets. Refer to Fig. IH 450 or 451. Differential carrier bearings on the Cub are of the adjustable, taper roller type; whereas, the A and B series carrier bearings are of the non-adjustable, ball type.

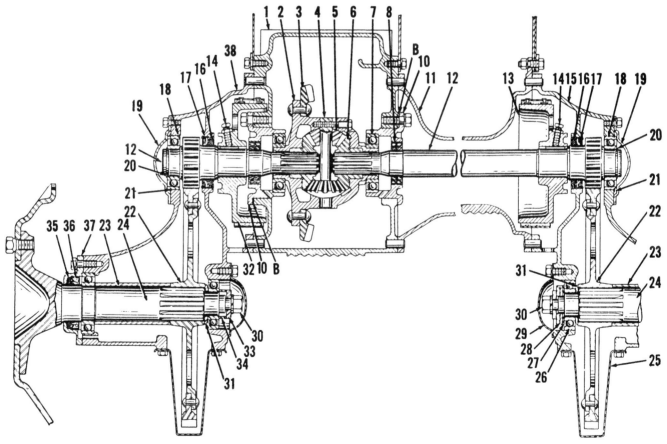

Fig. IH450—Sectional view of series A and B main drive bevel ring gear, differential and final drive assembly. The Cub construction is similar.

B. Shims	6. Differential gear	13. Brake drum	20. Snap ring	26. Bearing	32. Brake drum
1. Transmission case	7. Bearing	14. Set screw	21. Snap ring	27. Spacer	33. Washer
2. Differential case	8. Oil seal	15. Rear axle housing	22. Bull gear	28. Spacer	34. Spacer
3. Main drive bevel ring gear	9. Oil seal	16. Felt seal	23. Spacer	29. Bearing cap	35. Felt seal
4. Pinion shaft	10. Bearing retainer	17. Oil seal	24. Rear axle	30. Cap screw	36. Oil seal
5. Differential pinion	11. Differential shaft housing	18. Bearing	25. Rear axle housing pan	31. Spacer	37. Bearing retainer
	12. Differential shaft	19. Bearing cap			38. Rear axle housing

453. R&R AND OVERHAUL. To remove the differential assembly, first detach the final drive and differential shaft housing assemblies and the rear cover or power take-off and/or belt pulley attachment from the transmission case. Remove the differential bearing retainers (10—Fig. IH450), or (7—Fig. IH451) and lift the main drive bevel ring gear and differential unit from rear of transmission case. Note: The differential carrier bearings (7 —Fig. IH450) or (3 & 4—Fig. IH451) can be renewed at this time.

To disassemble the differential, remove the pinion shaft lock pin or retainer screw from differential case; then, remove pinion shaft, pinions and side gears.

To reassemble, reverse the disassembly procedure and reinstall unit in transmission case. On the Cub, shims (B) located between differential bearing retainers and transmission case control differential carrier bearing adjustment. Adjust bearings to provide free rotation with zero end play; then transfer shims from one retainer to the other to obtain a backlash of 0.003-0.005 between the bevel gears. On the A and B series, the differential carrier bearings are non-adjustable, but shims (B—Fig. IH450) should be arranged to provide a bevel gear backlash of 0.005-0.007.

Refer to paragraph 301 for method of adjusting tooth contact (mesh pattern) of the main drive bevel gears.

Model C-Super C

Differential unit is of the four pinion type enclosed in a two piece case which is mounted behind a dividing wall in the transmission case as shown in Fig. IH454. The bolts holding the two halves of the differential case together also secure the main drive bevel ring gear to the case. The carrier bearings are adjustable.

455. R&R AND OVERHAUL. To remove the differential assembly from tractor, drain rear frame (transmission case), and proceed as follows: Remove the final drive bull gears as outlined in paragraph 469, both bull pinions and bull pinion shaft bearing cages and lift the differential and bevel ring gear assembly from the case. Note: The differential carrier bearing cups can be removed from the bull pinion shaft bearing cages and the bearing cones can be removed from the differential case halves at this time.

456. To disassemble the unit, remove the assembly bolts, and separate the two halves. Differential assembly case bolt nuts are of the self-locking type. The shims (B—Fig. IH455), which are located between the bull pinion shaft bearing cages and the transmission case, control the differential carrier bearing adjustment and the backlash between the main drive bevel gears. Adjust the carrier bearings to a slight pre-load; then, transfer shims from

one retainer to the other to obtain a bevel gear backlash of 0.008-0.010.

Refer to paragraph 301 for method of adjusting the tooth contact (mesh pattern) of the main drive bevel gears.

Fig. IH454—Top view of Super C transmission case, showing installation of the transmission shafts, differential and bull gears. Model C construction is similar.

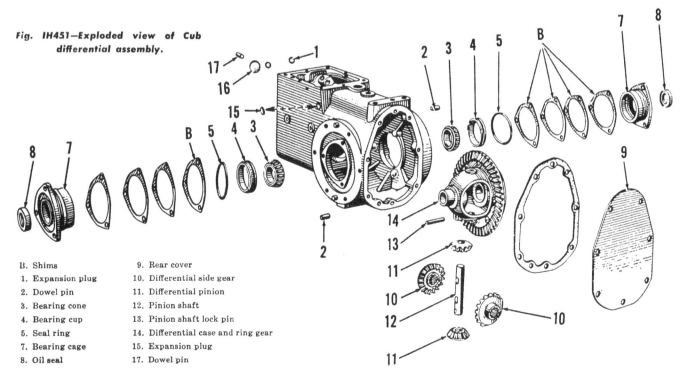

Fig. IH451—Exploded view of Cub differential assembly.

B. Shims
1. Expansion plug
2. Dowel pin
3. Bearing cone
4. Bearing cup
5. Seal ring
7. Bearing cage
8. Oil seal
9. Rear cover
10. Differential side gear
11. Differential pinion
12. Pinion shaft
13. Pinion shaft lock pin
14. Differential case and ring gear
15. Expansion plug
17. Dowel pin

Series H-M-4-6

Differential unit is of the four pinion open case type mounted back of a dividing wall in the rear frame (transmission case) as shown in Fig. IH457. The differential case halves are held together by bolts which also retain the bevel ring gear. The bearings supporting the differential are of the non-adjustable ball type.

458. R&R AND OVERHAUL. To remove the differential and final drive bull pinions without disturbing the transmission, proceed as follows: Remove rear frame (transmission case) cover and bull gears as outlined in paragraph 474. Remove brake drums. Remove cap screws retaining differential bearing cages to transmission case, remove the bull pinion shafts and bearing cages and lift the differential assembly from tractor. Note: The differential carrier bearings can be renewed at this time.

When reinstalling differential unit, assemble bull pinion shaft (10) to right hand bearing cage (11) and install the unit. Lower differential unit into rear frame and enter same over splines of bull (spur) pinion shaft. Install opposite bearing cage and bull pinion unit.

The differential carrier bearings are non-adjustable, but bearing cage-to-rear frame shims (B) are provided to position the main drive bevel ring gear for proper backlash in relation to the bevel pinion. Desired backlash of 0.008-0.012 is obtained by removing shim or shims (B) from one bearing cage (11) and inserting them under the other bearing cage. Refer to paragraphs 300 and 301 for method of adjusting tooth contact (mesh pattern) of main drive bevel gears.

Series 9

Differential unit is of the four pinion open case type mounted back of a dividing wall in the rear frame (transmission case). The differential case halves are held together by bolts which also retain the main drive bevel ring gear as shown in Fig. IH 460. The bearings which support the differential are of the non-adjustable ball type.

460. R & R AND OVERHAUL. To remove the differential unit from the rear frame (transmission case) it is necessary to first disconnect the rear axle housing as outlined in paragraph 480. With the rear axle housing detached from the rear frame, proceed as follows: Remove the brake drums and cap screws retaining differential bearing cages (8) to rear frame. Remove bearing cages and bull pinion units (11) as an assembly by using cage cap screws as a puller. Common practice is to place a one inch board under differential ring gear when removing and reinstalling assembly.

Bull pinion and differential shaft units can be bumped out of the bearing cages. Note: Differential carrier bearings can be renewed at this time. To prevent damage to rubber sealing rings (14) when reinstalling differential bearing cages, coat sealing rings with a thick soap solution.

On rice field models, install seal (9) with lips facing away from tractor centerline. On other models, seal lips face toward tractor centerline.

Reinstall differential assembly using care to install the bearing cages with the lubricating hole facing up. Shims (B) located between bearing cages and rear frame are used to adjust the tooth backlash of main drive bevel gears. Differential should rotate freely, with a minimum amount of end play. Desired backlash of 0.010-0.012 is obtained by removing shims (B) from one bearing cage and inserting them under the opposite bearing cage. Refer to paragraphs 300 and 301 for method of adjusting tooth contact (mesh pattern) of main drive bevel gears.

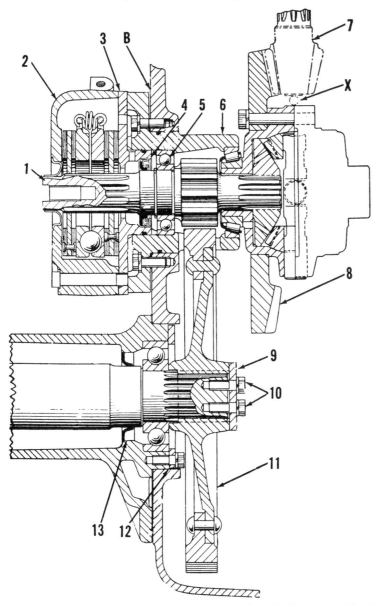

Fig. IH455—Sectional view of Super C differential and final drive assembly. Rod (X) is used to set trial position of the main drive bevel pinion. Model C construction is similar except band type brakes are used and seal (4) is carried in the bull pinion shaft bearing quill.

B. Shims	3. Bearing retainer	7. Bevel pinion
1. Bull pinion and	4. Oil seal	8. Bevel ring gear	11. Bull gear
brake shaft	5. Bearing	9. Washer	12. Retainer plate
2. Brake housing	6. Bearing quill	10. Cap screws	13. Oil seal

FINAL DRIVE

Series A-B

As treated in this section the final drive is assumed to be the differential shafts (bull pinion shafts) and housings and the rear axle housings containing drive (bull) gears and axles and brake assemblies. Refer to Fig. 450.

462. DIFFERENTIAL (BULL PINION) SHAFTS. Removal of left differential shaft requires removal of left rear wheel. On series A, remove cap screws retaining left rear wheel axle housing (38—Fig. IH450) to transmission case and remove rear axle housing assembly. On models B and BN, remove cap screws retaining left differential shaft housing (interposed between transmission and rear axle housing) to transmission case and remove rear wheel axle housing assembly. Remove brake drum.

Remove cap screws retaining differential shaft bearing cover to the housing and bump shaft on inner end and out of housing.

Removal of right differential shaft requires removal of right rear wheel. Remove cap screws retaining differential shaft housing (11) to transmission case and remove as an assembly with rear wheel axle housing. Separate differential shaft housing from rear wheel axle housing (leaving the platform and seat assembled to it). Remove brake drum and cap screws retaining differential shaft bearing cover (19) to rear axle housing and bump shaft on inner end and out of housing.

463. R&R WHEEL AXLE SHAFT AND/OR BULL GEAR. To remove either rear wheel axle (24—Fig. IH 450) or drive (bull) gear (22), proceed as follows: Block rear end of tractor to raise wheel off of ground. Remove wheel and cap screws retaining outer bearing retainer (37) and inner bearing cover (29) or planter drive to the housing. Remove bearing retainer cap screw (30) and washer. Remove rear axle housing pan (25).

Bump shaft on inner end and out of inner bearing, spacer, drive gear and rear axle housing.

Rear axle shaft bearings are nonadjustable. Install rear axle outer seal with lip toward the bearing.

Model Cub

465. The final drive (differential shaft and rear axle shaft) construction is similar to that used on the models A and AV. The rear wheel axle bearings on the Cub are adjustable. Adjustment is made with rear wheel axle nut and is correct when axle shaft has zero end play but rotates freely.

To adjust rear axle bearings, block rear end of tractor to raise wheel off ground. Remove cap screws retaining wheel axle nut oil seal retainer to inner end of axle housing and remove retainer. Remove cotter key from axle shaft and rotate nut to adjust the axle bearings. Refer to paragraphs 462 and 463 for method of R & R of differential (bull pinion) shafts and rear wheel axle shafts.

Series C

The final drive consists of two bull pinions and integral brake shafts and two bull gears which are splined to the inner ends of the rear axle shafts. Each axle shaft is carried in a sleeve which is bolted to the side of the transmission case.

468. R&R BULL PINION (DIFFERENTIAL) SHAFTS. To remove either bull pinion and integral shaft, first remove the respective brake unit. On Model C, extract seal and snap ring from the bull pinion shaft bearing cage and withdraw the bull pinion

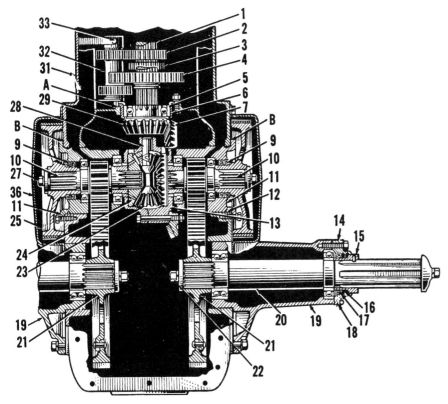

Fig. IH457—Sectional view of series 4 and 6 final drive as used on tractors with band type brakes. Series H and M are similar. When disc brakes are used, the bull pinion shaft and bearing quill are of different construction; refer to Fig. IH458. Refer to Fig. IH463 for high clearance models.

A. Shims	7. Bevel ring gear
B. Shims	9. Oil seal
1. Countershaft	10. Bull pinion
2. First gear	11. Bearing cage
3. Second gear	12. Seal ring
4. First and reverse gear	13. Differential case
5. Bearing	14. Bearing retainer
6. Bearing cage	15. Bearing spacer (4 and 6 series only)
16. Felt seal	25. Brake drum
17. Oil seal	27. Cap screw
18. Lubricator	28. Differential spider
19. Rear axle carrier	29. Main drive bevel pinion
20. Wheel axle shaft	31. Rear frame
21. Bull gear	32. Reverse idler gear
22. Cap screw	33. Reverse idler shaft retaining screw
23. Differential pinion	
24. Side gear	

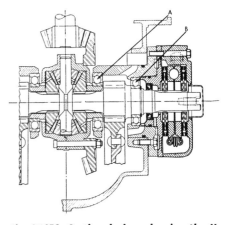

Fig. IH458—Sectional view showing the H, M, 4 and 6 series bull pinion shaft as used on tractors with disc type brakes. The shaft outer bearing (B) and seal, and the bull pinion shaft can be removed without disturbing the transmission top cover.

shaft. On the Super C, remove the bull pinion shaft bearing retainer plate (inner brake drum) (3—Fig. IH 455) and withdraw the bull pinion shaft and bearing from the bearing cage.

469. R&R BULL GEAR. To remove the left bull gear, drain the transmission case and remove the transmission case top cover. Remove the rear wheel and cap screws retaining bull gear to inner end of the wheel axle shaft. Remove the axle carrier retaining cap screws and withdraw axle and carrier assembly from the transmission case. Lift the bull gear

from the case. To remove the right bull gear, follow the same procedure as for the left, except it is usually necessary to remove the belt pulley and power take-off unit and remove the belt pulley oiler tube.

470. R&R WHEEL AXLE SHAFT. To remove either wheel axle shaft, first remove the respective bull gear as outlined in the preceding paragraph and proceed as follows: Remove cap from outer end of the axle sleeve and bump shaft on inner end and out of the sleeve. The axle inner bearing and seal can be renewed after removing the bearing retainer ring from

the inner end of the sleeve. The procedure for renewing the outer seals and bearing are evident. Install seals with lips of same facing the bull gear.

Series H-M-4-6

As treated in this section, the final drive will include the bull pinions and integral shafts, both bull gears and both rear wheel axle and sleeve assemblies.

473. R&R BULL PINION (DIFFERENTIAL) SHAFTS. To remove either bull pinion on models equipped with band type brakes, support rear of tractor and remove the respective bull gear as outlined in paragraph 474. Remove brake drum and cap screws retaining the differential bearing cage to the transmission case. The bearing cage and bull pinion shaft can be withdrawn as an assembly by using the bearing cage cap screws as a puller.

To remove either bull pinion on models equipped with disc type brakes, remove the respective brake unit, remove the bull pinion shaft bearing retainer plate (inner brake drum) and withdraw the bull pinion shaft and bearing from the bearing cage.

474. R&R BULL GEAR. Remove rear frame top (transmission and final drive) cover. Remove rear wheel and cap screw (22—Fig. IH457 or 462) or (36—Fig. IH463) retaining bull gear on inner end of wheel axle or bull sprocket shaft. Remove cap screws retaining rear axle or bull sprocket carrier or sleeve to rear frame. Withdraw carrier and shaft as a unit from the rear frame. (On high clearance models withdraw carrier and one final drive unit as an assembly.) Bull gear can now be removed from rear frame.

475. R&R WHEEL AXLE SHAFT. To renew the wheel axle shaft or bearings on Series 4 and 6, first detach bull gear as described in paragraph 474. With carrier off tractor, remove spacer and pin (15—Fig. IH457) from outer end of shaft then remove outer bearing retainer (14). Bump outer end of axle shaft. This will dislodge the shaft and inner ball bearing. The outer bearing can now be drifted from the carrier using a long drift against the outer race.

475A. To renew wheel axle shaft or bearing on the H and M Series (except high clearance models), first detach bull gear as described in paragraph 474. With carrier off tractor, remove outer bearing retainer and remove shaft from carrier by driving on its inner end. Refer to Fig. IH462.

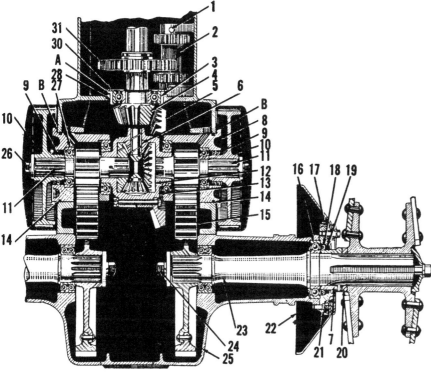

Fig. IH460—Sectional view of series 9 final drive assembly. The rear axle housing assembly can be separated from the transmission case as shown in Fig. IH464. Late models have disc type brakes and the bull pinion shaft installation is different.

A. Shims	6. Differential pinion	14. Seal ring
B. Shims	7. Bearing spacer	15. Bull gear
1. Reverse idler shaft retaining screw	8. Bearing cage	16. Bearing
2. Reverse idler	9. Oil seal	17. Lubricator
3. Bearing	10. Brake drum	18. Oil seal
4. Main drive bevel pinion	11. Bull pinion and shaft	19. Felt seal
5. Main drive bevel ring gear	12. Side gear	20. Spacer
	13. Differential case	21. Bearing retainer
		22. Fender support

23. Rear wheel axle shaft
24. Cap screw
25. Rear housing
26. Cap screw
27. Bearing
28. Bearing cage
30. Bearing cage cover
31. First and reverse gear

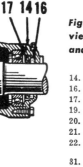

Fig. IH462 — Sectional view, showing rear axle and bull gear installation on series H and M.

14. Bearing retainer
16. Felt seal
17. Oil seal
19. Rear axle carrier
20. Rear axle
21. Bull gear
22. Bull gear retaining cap screw
31. Rear frame
34. Bearing retainer

475B. To remove wheel axle shaft or sprocket on high clearance models of the H and M series, proceed as follows: After draining the lubricant, remove housing pan (31 — Fig. IH463) and the connecting link from drive chain (25). It is not necessary to remove the drive chain. If removal of drive chain is desired, fasten a piece of flexible wire to one end and thread chain off of sprocket (20). The wire is used to help install and thread the chain over the top sprocket.

Remove inner and outer wheel axle bearing retainers (27 and 32) and inner bearing cap screw (33). Bump wheel axle shaft (28) on inner end and out of sprocket while bucking up the sprocket (44).

When reinstalling axle shaft, install spacer (42) between inner bearing (34) and sprocket hub with wide face of spacer nearest the sprocket hub.

478. R&R SPROCKET (UPPER) SHAFT. To remove the upper sprocket shaft (17—Fig. IH463) on high clearance models of the H and M series, proceed as follows: Remove bull gear as outlined in paragraph 474. Remove drive chain as outlined in first part of paragraph 475B. Remove outer bearing cap (24) and cap screw (23) retaining bearing (22) to shaft. Remove cap screws retaining sprocket shaft carrier (18) to final drive housing (21) and remove carrier (18). Bump sprocket shaft out of sprocket and withdraw the sprocket.

Series 9

480. R & R REAR AXLE HOUSING ASSEMBLY. To remove rear axle housing from the rear frame (transmission case) as shown in Fig. IH464, remove fenders, fender support braces and seat. Remove power take-off and disconnect swinging drawbar if tractor is so equipped. Take out bolts retaining platform to rear frame and disconnect brake lock and pedal return spring.

Support tractor under rear frame and drive wooden wedges between ends of buffer spring and front axle to prevent tractor from tilting when rear axle is detached. Remove rear axle housing to rear frame stud nuts and cap screws and separate rear axle housing from rear frame. Rear axle housing is not balanced on the rear axles so caution must be used when separating from rear frame to prevent the rear face of housing from tilting downward.

481. R&R BULL PINION SHAFTS. To remove either bull pinion, support rear of tractor and remove the rear axle housing assembly as outlined in the preceding paragraph. Remove the brake drum and cap screws retaining the differential bearing cage to transmission case. Remove the bearing cages and bull pinion units as an assembly by using the cage retaining cap screws as a puller, threaded into holes provided in the bearing cage flange.

Install the double opposed seal in the shaft bearing cage with lips facing away from differential on rice field models; toward differential on other models.

482. R&R WHEEL AXLE OR BULL GEAR. Axle shaft (23—Fig. IH460) or bull gear (15) can be removed after disconnecting rear axle housing from rear frame, as outlined in paragraph 480. To remove bull gear proceed as follows: Let rear of housing tilt down so opening will be up. Block up housing and remove wheel. Remove cap screw (24—Fig. IH460) and thrust washer retaining bull gear to inner end of shaft. Remove cap screws retaining bearing retainer (21) to axle housing. Replace cap screw in inner end of axle and press or bump shaft out of rear axle housing by applying pressure on inner end of shaft and withdraw shaft out of bull gear and housing.

On models W9 and WD9, install the oil seal (18—Fig. IH460) with lip facing toward the bearing. On models WR9 and WDR9, install seal with lip facing away from the bearing.

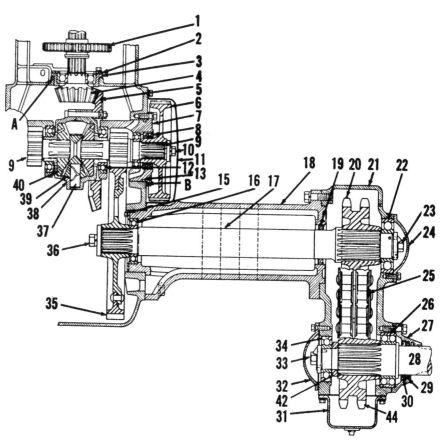

Fig. IH463—Sectional view of series H and M high clearance rear axle, bull gear and differential assembly as used on models with band type brakes. Models using disc type brakes are similar except for the bull pinion installation; refer to Fig. IH458.

A. Shims	11. Brake cover	27. Bearing retainer
B. Shims	12. Bearing	28. Rear axle
1. First and reverse gear	13. Seal ring	29. Felt seal
2. Bearing retainer	15. Bearing retainer	30. Oil seal
3. Bearing cage	16. Bearing	31. Axle housing pan
4. Main drive bevel pinion	17. Drive sprocket shaft	32. Bearing cap
5. Main drive bevel ring gear	18. Housing carrier	34. Bearing
6. Brake drum	19. Oil seal	35. Bull gear
7. Bearing cage	20. Rear axle drive sprocket	36. Bull gear retaining screw
8. Oil seals	21. Rear axle housing	37. Spider
9. Bull pinion and brake shaft	22. Bearing	38. Differential case
10. Drum retaining screw	24. Rear axle cap	39. Differential pinion
	25. Rear axle drive chain	40. Differential side gear
	26. Bearing	42. Spacer
		44. Rear axle sprocket

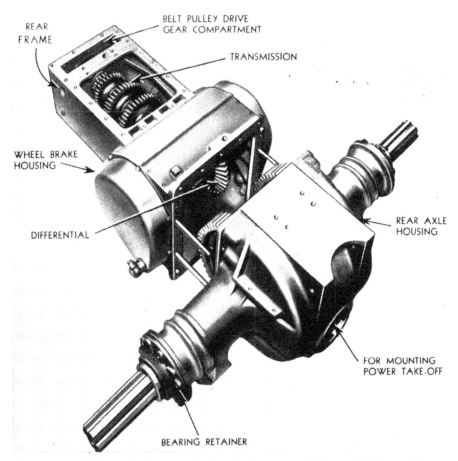

Fig. IH464—Splitting series 9 rear axle housing from transmission and differential case.

Series C-H-M-4-6-9 (Band Type)

494. Brakes are band type, similar to those described in paragraph 490. Brake drums are mounted on the outer ends of bull pinion shafts, enclosed by individual housings, and the brake cross shaft passes through the rear frame (transmission case). Models O4, OS4, O6, OS6 and OSD6, do not have a cross shaft, but have an individual brake lock control rod on each side of the seat support. Procedure for removing the brake band is evident after an examination of the unit.

495. **ADJUST BRAKES.** Brake pedals should have a free travel of 1⅜ to 1½ inches, measured from the pedal stop against transmission cover. To adjust, first turn the set screw (A—Figs. IH466, IH467 and IH468) on the bottom of each brake housing in as far as it will go, then back out ½ turn or until drag is removed. Obtain the pedal free travel by changing the length of the brake rod (1). Equalize the brakes by loosening the adjustment on the tight brake.

Series C-H-M-4-6-9 (Disc Type)

497. Brakes are of the double disc, self energizing type, which are splined

BRAKES

Series A-B-Cub

490. Brakes are external band type contracting on drums mounted on differential (bull pinion) shafts and enclosed by the rear wheel axle housings. Brake pedals have an interlock and bands are connected to pedals by brake rods and a cross shaft mounted underneath clutch housing (on the Cub this cross shaft is mounted in the clutch housing). Relining of the brakes requires removal of the final drive assembly and differential shaft which gives free access to brake bands.

491. **ADJUST.** Adjustment is accomplished by changing the length of the brake rod (1—Fig. IH465) by means of the adjustable clevis (2) at pedal end of rod until each pedal has a free movement of approximately 1 inch as measured at top of pedal. To equalize the brakes, jack up rear of tractor and operate in high gear. Apply brakes. If both wheels do not slow down together, loosen the adjustment on the side which stops first until both wheels slow down at same time when brakes are applied.

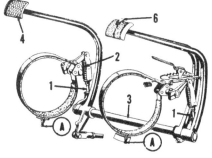

Fig. IH 467—Typical band type brake assembly for series 9 and models W4 and W6.

A. Adjusting screw
1. Brake rod
2. Spring
3. Pedal cross-shaft
4. Clutch pedal
6. Pedal interlock plate

Fig. IH 468—Typical band type brakes for models O4, OS4, O6, OS6 & ODS6.

A. Adjusting screw
1. Brake rod
2. Spring
5. Pedal return spring

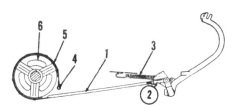

Fig. IH 465—Typical brake assembly for series A, B and Cub.

1. Brake rod
2. Adjusting clevis
3. Pedal return spring
4. Brake band anchor
5. Brake band
6. Brake drum

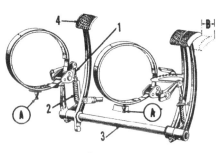

Fig. IH 466—Typical band type brake assembly for series H and M.

A. Adjusting screw
B. Pedal free travel
1. Brake rod
2. Brake compression spring
3. Pedal cross-shaft
4. Clutch pedal

to the outer ends of the bull pinion and integral brake shaft. The moulded linings are riveted to the brake discs. Procedure for removing the lined discs is evident after an examination of the unit.

498. ADJUSTMENT. To adjust the brakes, loosen jam nut (5—Fig. IH 470) and turn rod (1) either way as required to obtain the correct pedal free travel of $1\frac{1}{8}$-$1\frac{3}{8}$ inches ($2\frac{1}{8}$ inches on series 9). Brakes can be equalized by loosening the tight brake. When adjustment is completed, tighten jam nut (5). If for any reason the brake operating rod assembly has been disassembled, spring (2) should be adjusted to a pre-load dimension of $1\frac{3}{16}$ inch. Assemble the component parts as shown and turn nut (3) until the correct dimension is obtained, and lock same by tightening jam nut (4).

Fig. IH 470 — Cut-away view showing a typical disc type brake.

1. Adjusting rod
2. Spring
3, 4 & 5. Nuts
6. Housing
7. Lined disc
8. Actuating disc
9. Spring
10. Bearing retainer (inner brake drum)
11. Bearing quill
12. Ball

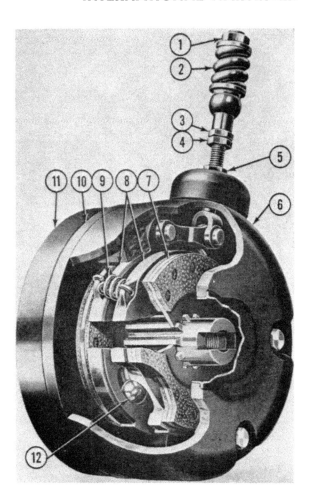

BELT PULLEY AND POWER TAKE-OFF

Series A-B-C

499. The belt pulley and power take-off assembly shown in Fig. IH 475 is driven by the shaft (11) and coupling (12) which engage the transmission spline shaft (20). The forward end of drive shaft (11) is supported in a pilot bushing (21) in the end of transmission spline shaft. Disassembly and reassembly of the unit is apparent after studying the illustration. The double leather oil seal (16) on the take-off shaft is installed with lips facing inward and the oil seal (5) in belt pulley shaft outer bearing cage should also be installed with lip facing inward. Shims (X) and (Y) are provided for proper adjustment of backlash and tooth contact of the two bevel gears. Adjust gears so that heels are in register and backlash is from 0.004-0.006 (I&T recommended).

Note: Earlier belt pulley and power take-off attachments (48153D and 48104D) had the shims (X) between power take-off shaft rear bearing cage and the belt pulley housing in which

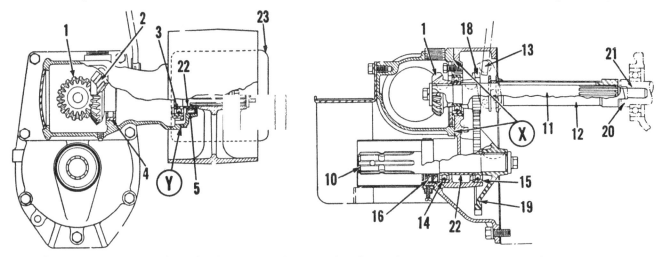

Fig. IH 475—Cut-a-way views showing the combination belt pulley and power take-off unit as used on series A-B-C.

X. Shims	3. Pulley shaft outer bearing	10. PTO shaft	15. PTO shaft inner bearing	20. Transmission spline shaft
Y. Shims		11. Drive shaft	16. PTO shaft oil seal	21. Drive shaft bushing
1. Bevel drive gear	4. Inner bearing	12. Shifter coupling	18. Drive shaft pinion	22. Bearing spacer
2. Pulley shaft driven gear	5. Pulley shaft oil seal	14. PTO shaft outer bearing	19. PTO shaft drive gear	23. Belt pulley guard

the cage was then mounted. On the front end of the power take-off shaft an oil slinger also was installed. (Non-standard 1⅛ inch spline power take-off attachment can be converted to standard 1⅜ inch power take-off spline shaft with IH package 66191D.)

Cub

500. The belt pulley is driven by the power take-off shaft which mounts in the transmission case. The complete assembly is shown installed in Fig. IH432. The forward end of the power take-off shaft (10) is supported in a bushing in rear of the transmission spline shaft and is driven by the spline shaft through a shifter coupling (11). The belt pulley housing and support, bolt to the rear of the transmission case and an internally splined sleeve in the assembly fits over the splines of the power take-off shaft. A bevel gear (3) is keyed to this sleeve and meshes with a bevel gear (2) on the belt pulley shaft.

The combination felt and spring loaded leather seal (7) for power take-off shaft is assembled with felt to rear and lip of leather seal facing forward. The spring loaded leather seals in pulley housing support and for belt pulley shaft are installed with lips facing toward the bevel gears. Shims (X) between pulley housing and support and shims between pulley shaft outer bearing cage and pulley housing are provided for proper adjust-

ment of backlash and tooth contact of the two bevel gears. Adjust gears so that heels are in register and backlash is from 0.004-0.006 (I&T recommended).

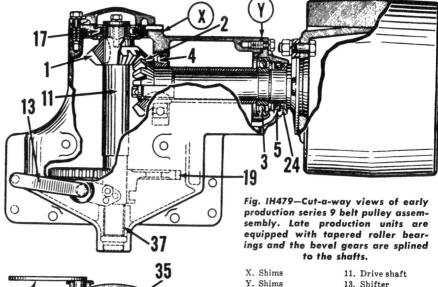

Fig. IH479—Cut-a-way views of early production series 9 belt pulley assemsembly. Late production units are equipped with tapered roller bearings and the bevel gears are splined to the shafts.

X. Shims	11. Drive shaft
Y. Shims	13. Shifter
1. Bevel drive gear	17. Drive shaft
2. Bevel driven gear	bearing
3. Pulley shaft outer	19. Drive gear
bearing	24. Felt washer
4. Pulley shaft inner	34. Pulley housing
bearing	35. Shifter poppet
5. Pulley shaft oil	37. Drive shaft
seal	bushing

Series H-M-4-6-9

505. **BELT PULLEY.** The belt pulley unit is mounted on transmission cover and driven by the transmission. Procedure for disassembly and reas-

sembly is apparent after studying Fig. IH477 for the H, M, 4 and 6 series, or Fig. IH479 for the 9 series. Shims (X) and (Y) are provided for adjustment of backlash and tooth contact of the bevel gears. Gear backlash should be 0.008-0.010. Leather seal (5) should be installed with lip facing bearing. The felt seal (24) in both cases should be installed outside of the leather seal. Soak this felt seal in oil before installing.

508. **NON-CONTINUOUS POWER TAKE-OFF.** Fig. IH483 shows the power take-off unit for the H, M, 4 and 6 series; Fig. IH485 is for the 9 series after 14400. Power take-off shown in Fig. IH486 is the assembly used in 9 series, 501 to 14400. In all cases, the forward end of the power take-off shaft is supported in pilot bushing (21) in end of the transmission countershaft.

Disassembly and reassembly is apparent after studying the illustrations. The double leather seal (16) should be installed with the lips facing inward. Note: Early H and M models' transmissions which have a notched-end countershaft must have the later splined-end-type countershaft (59353-DX or 60123DX for H and 59354DX or 60125DX for M) installed before the power take-off unit can be attached.

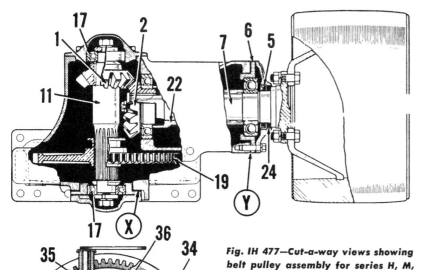

Fig. IH 477—Cut-a-way views showing belt pulley assembly for series H, M, 4 and 6.

X. Shims	11. Drive shaft
Y. Shims	17. Drive shaft
1. Bevel drive gear	bearing
2. Bevel driven gear	19. Drive gear
5. Pulley shaft oil	22. Spacer
seal	24. Felt washer
6. Outer bearing	34. Pulley housing
cage	35. Shifter poppet
7. Pulley shaft	36. Shifter stop

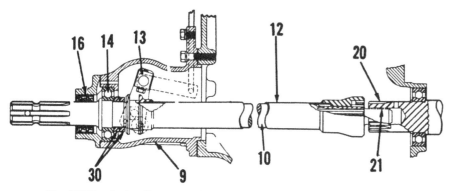

Fig. IH483—Series H. M. 4 and 6 non-continuous power take-off assembly.

9. PTO housing
10. PTO shaft
12. Shifter coupling
13. Shifter

14. PTO shaft outer bearing
16. PTO shaft oil seal

20. Transmission countershaft
21. Pilot bushing
30. Double nut

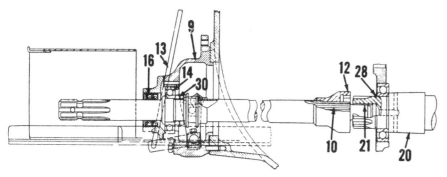

Fig. IH485—Power take-off assembly as used on series 9 after 14400.

9. PTO housing
10. PTO shaft
12. Shifter coupling

14. Outer bearing
16. PTO shaft oil seal
20. Transmission countershaft

21. Pilot bushing
28. Snap ring
30. Double nut

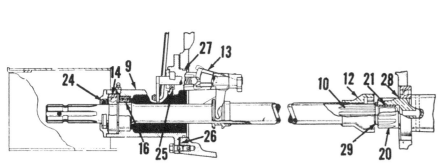

Fig. IH486—Power take-off assembly as used on series 9 (501-14400).

9. PTO housing
10. PTO shaft
12. Shifter coupling
13. Shifter
14. Outer bearing

16. PTO shaft oil seal
20. Transmission countershaft
21. Pilot bushing

24. Felt washer
26. Housing shims
27. Rubber ring
28. Snap ring
29. Thrust washer

508A. INDEPENDENT POWER TAKE-OFF. The occasion for overhauling the complete power take-off system will be infrequent. Usually, any failed or worn part will be so positioned that localized repairs can be accomplished. The subsequent paragraphs will be outlined on the basis of local repairs. If a complete overhaul is required, a combination of the appropriate following paragraphs can be used.

508B. ADJUST REACTOR BANDS. To adjust the reactor bands, remove the adjusting screw cover, loosen lock nuts (N—Fig. IH486A) and back off the adjusting screws (M) approximately four turns. Hold the operating lever so that punch marked dot on pawl is aligned with the punch dot on quadrant as shown.

Turn the adjusting screws in (clockwise) until screws are reasonably tight and set the lock nuts finger tight. Move the operating lever back and forth several times; then, back-off the adjusting screws approximately one turn and tighten the lock nuts.

Using a 5/16 inch rod approximately 8 inches long as a lever in hole of pto shaft, check to make certain that shaft is free to turn only when punched dots on pawl and quadrant are aligned. The shaft should not turn with lever in any other position.

If shaft will not turn when punched dots are aligned, the adjusting screws are too tight.

Tighten lock nuts when adjustment is complete.

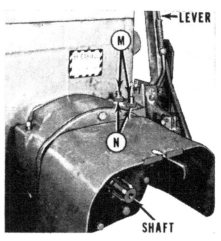

Fig. IH486A—When adjusting the independent power take-off reacter bands, punched mark on pawl should be aligned with the punch mark on the quadrant.

508C. RENEW REACTOR BANDS. To renew the reactor bands (34—Fig. IH487), remove the pto shaft guard and the band adjusting screw cover. Unbolt bearing retainer (47) from the housing cover and remove the retainer. Unlock the pto shaft nut (45), place the operating lever in the forward position to prevent shaft from turning and remove nut (45). Remove the cap screws retaining the rear cover to the housing, place operating lever in the neutral position and turn the pto shaft to align threaded holes in the rear drum (38) with the unthreaded holes in the rear cover.

Using OTC puller ED-3262 or equivalent as shown in Fig. IH487B, remove the rear drum and cover. Loosen the band adjusting screws and remove bands. Inspect bands for distortion, lining wear and for looseness of strut pins.

When reassembling, adjust the bands as outlined in paragraph 508B.

508D. OVERHAUL PLANET GEARS, SUN GEARS AND SHAFTS. To overhaul the pto rear unit, first drain transmission, then remove the reactor bands as outlined in paragraph 508C. Unbolt and remove the unit from the tractor rear frame.

With the unit on bench, remove cap screws securing bearing cage (12—Fig. IH487 or 487A) to housing and remove bearing cage with ring gear and shaft. Remove the front bearing retainer (16) and seal (17). Remove snap ring (14) and press the ring gear and shaft from bearing (13). The front bearing (13) can be removed from cage at this time. To remove the ring gear

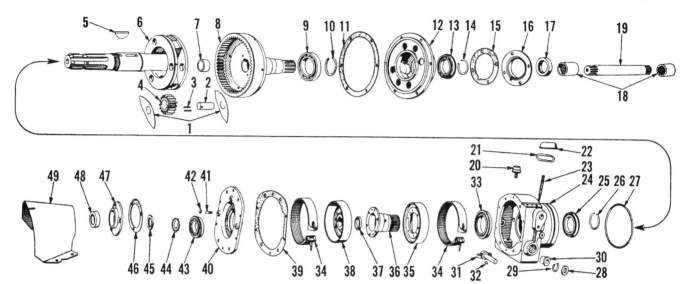

Fig. IH487—Exploded view of the rear section of the independent power take-off. Planetary gears (4) are available in sets only.

1. Thrust washers	9. Drive shaft rear bearing	16. Bearing retainer	32. Key
2. Planet gear shaft	10. Snap ring	17. Oil seal	33. Bearing
3. Needle bearings	11. Gasket	18. Couplings	34. Brake band
4. Planet gear	12. Bearing cage	19. Coupling shaft	35. Brake drum
5. Key	13. Drive shaft front bearing	20. Breather	36. Sun gear
6. Planet carrier and pto shaft	14. Snap ring	21. Gasket	37. Spacer
7. Needle bearing	15. Gasket	22. Anchor bolt cover	38. Creeper drum
8. Ring gear and shaft		23. Bolt	39. Gasket
		24. Housing	40. Housing cover
		25. Bearing	41. Stud
		26. Snap ring	42. Nut
		27. Seal ring	43. Bearing
		28. Oil seal	44. Lock washer
		29. Snap ring	45. Nut
		30. Bushing	46. Gasket
		31. Lever	47. Bearing retainer
			48. Oil seal
			49. Shaft guard

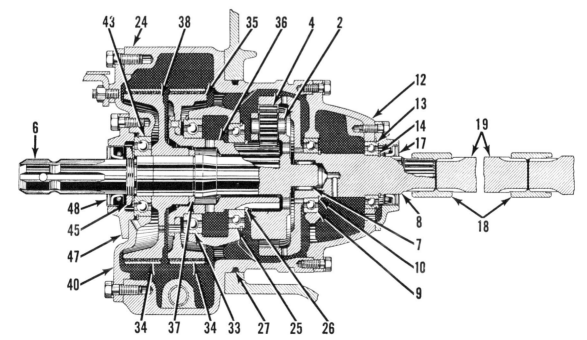

Fig. IH487A—Sectional view of the independent power take-off rear unit. Refer to legend for Fig. IH487.

rear bearing (9), remove snap ring (10) and using a punch through the two holes in the ring gear hub, bump bearing from shaft. Inspect needle roller pilot bearing (7). If the bearing is damaged, it can be removed, using

Fig. IH487B—Using OTC puller to remove cover and rear drum from the independent power take-off rear unit.

Fig. IH487C—Removing needle pilot bearing from ring gear.

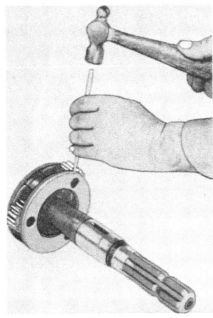

Fig. IH488—Removing the Esna roll pins which secure the planet gear shafts in the planet carrier.

a suitable puller as shown in Fig. IH487C.

Withdraw the planet carrier and pto shaft from housing and mark the rear face of each planet gear so it can be installed in the same position. Using a punch as shown in Fig. IH488, remove the Esna roll pins which retain the planet gear shafts in the planet carrier and remove the shafts, gears, spacers and needle bearings. Remove snap ring (26—Fig. IH487 or 487A) from sun gear and press sun gear (36) and drum (38) from housing. Bearing (25) can be removed from housing at this time. If rear bearing (33) is damaged, use a punch through holes in hub and drift the bearing from the sun gear as shown in Fig. IH488A. Remove operating linkage from side of housing. To disassemble the spring retainer plug assembly, turn spring anchor

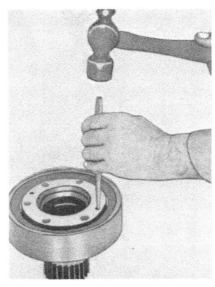

Fig. IH488A—Using a punch through holes of sun gear to remove rear bearing.

Fig. IH488B—Rear view of the tractor rear frame, showing the installation of the independent power take-off rear unit.

block clockwise to relieve spring pressure and remove snap ring. **Caution:** If the anchor bolt is broken or if spring tension cannot be relieved, use care when removing the snap ring.

Inspect all parts and renew any which are excessively worn. The planet gears are available only in sets as are the planet gear needle bearings and the gear shafts. Friction surface of drums should not be excessively worn.

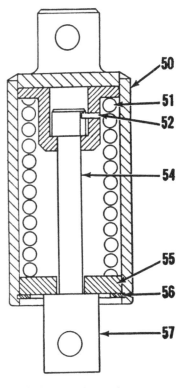

Fig. IH488C—Sectional view of spring retainer plug assembly. Refer to caution in text before disassembling.

50. Spring sleeve	55. Retainer plate
52. Roll pin	56. Snap ring
54. Anchor bolt	57. Anchor block

PTO REAR UNIT

When reassembling, reverse the removal procedure. OTC dummy shaft No ED-3258-1 is used to assemble the needle bearings in the planet gears and to install the planet gears, with thrust plates to the planet carrier. With the planet gear, thrust plates and dummy shaft in position in the planet carrier, push the dummy shaft out with the new shaft. Secure the planet gear shafts to the planet carrier with the Esna roll pins.

Adjust the reactor bands as outlined in paragraph 508B.

509. COUPLING SHAFT. To renew the pto coupling shaft (19—Fig. IH 487A), remove the complete pto rear unit (Fig. IH488B) and withdraw coupling shaft from rear frame.

509A. EXTENSION SHAFT. To remove the pto extension shaft (Fig. IH489), detach clutch housing from rear frame as outlined in paragraph 278. Remove the seasonal disconnect coupling from front of shaft. Unbolt the extension shaft front bearing retainer and cage from rear frame and withdraw the extension shaft assembly as shown in Fig. IH489A.

Remove the coupling shaft. Remove cap screw and strap which retains the extension shaft rear bushing carrier (Fig. IH489) in the rear frame and remove the bushing carrier. Bushing can be renewed if it is worn.

Reinstall the extension shaft by reversing the removal procedure.

509B. DRIVEN SHAFT & GEAR. Refer to Fig. IH489 .To remove the pto driven shaft and gear, first detach (split) engine from clutch housing as outlined in paragraph 214C and remove belt pulley unit.

Remove fuel tank, fuel tank support and air cleaner. Remove starting motor and the power take-off seasonal disconnect cover. Block-up rear frame, unbolt and remove clutch housing.

Remove the driven gear cover from bottom of clutch housing and the large pipe plug which is located directly in front of the driven shaft. Remove cap screw and washer retaining driven gear to shaft and remove snap ring from behind the driven shaft rear bearing. Using a brass drift, bump driven shaft rearward out of clutch housing and withdraw the driven gear. If the driven shaft front needle bearing is damaged, it can be renewed at this time. The needle bearing race on shaft can also be renewed if damaged. Remove the race with a brass drift to avoid damaging the shaft.

When reassembling, reverse the removal procedure and use sealing compound around pipe plug in front of driven gear.

509C. DRIVE SHAFT. Refer to Fig. IH489. To remove the driving shaft and integral gear, first detach (split) engine from clutch housing as outlined in paragraph 214C and remove the engine clutch release bearing and shaft. Unbolt the drive shaft front bearing cage and withdraw the drive shaft and bearing cage from clutch housing. The need and procedure for further disassembly is evident.

Fig. IH489A — Removing the independent power take-off extension shaft from rear frame.

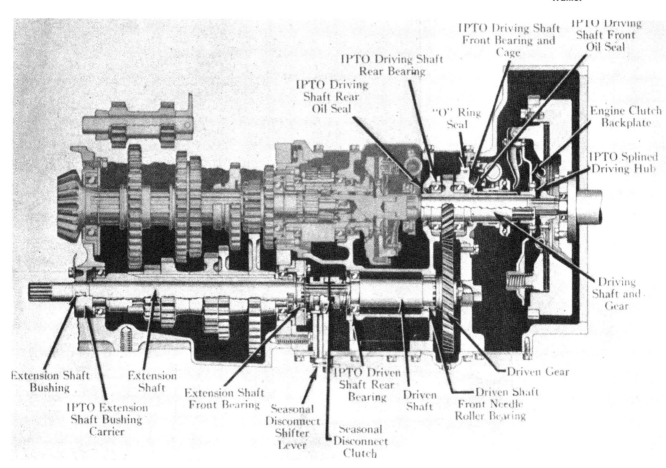

Fig. IH489—Sectional view of the clutch and transmission housing showing the installation of independent power take-off.

HYDRAULIC SYSTEM
(SERIES MTA & W6TA)

NOTE: The maintenance of absolute cleanliness of all parts is of utmost importance in the operation and servicing of the hydraulic system. Of equal importance is the avoidance of nicks or burrs on any of the working parts.

LUBRICATION

510. It is recommended that only IH Touch-Control Fluid be used in the hydraulic system. Reservoir is filled to the proper level when fluid is visible in the filler opening.

To bleed air from the hydraulic lines and remote control cylinder, loosen the stroke limit clamp on cylinder piston rod and move to yoke end so that piston can operate at maximum stroke. With remote cylinder in the retracted position, add fluid until reservoir is full. Start engine and operate at idle speed.

With the reservoir filler plug removed, operate the cylinder piston to its maximum stroke by moving the control lever back and forth about ten times to purge the system of trapped air. With the remote cylinder retracted, stop engine and refill the reservoir.

SYSTEM CHECKS

511. Before removing a faulty unit from tractor, it is advisable to first make a test to determine the cause of failure.

To check the hydraulic system pressure, insert a ½ inch pipe tee and a shut-off valve in series with the pump discharge hose, then install a pressure gauge, of at least 1500 psi capacity as shown in Fig. IH490A. Installation of a shut-off valve in the pump circuit will permit a separate check of the pump. With engine operating, hydraulic oil at operating temperature and shut-off valve fully opened, move the control lever to the raise position. Relief valve should unload at approximately 1000 psi for the Super W6TA, 750-1000 psi for the Super MTA.

If there is no pressure or if the pressure reading is low, slowly close the shut-off valve. Do not operate more than a few seconds with valve closed. If pressure fails to increase with the shut-off valve closed, the pump can be considered faulty. If the pressure does increase with the shut-off valve closed, check first for a leaking work cylinder, and then for a faulty relief valve.

The Super W6TA pressure relief valve is located externally on the reservoir, as shown in Fig. IH497, and can be removed after removing the relief valve cover (1). Relief spring has free length of 2 29/64 inches and should test 147 lb. at a length of 1 53/64 inches.

The Super MTA relief valve (17—Fig. IH493) is located in the control valves body. To remove the relief valve, it will be necessary to remove the oil reservoir and valves unit from tractor.

While making the foregoing check on the Super MTA tractor, the control rod should remain in the raise position until the work cylinder reaches the end of the stroke. After completing the lifting operation, the control rod should return automatically to the neutral position.

If the control rod does not remain in the raise position while lifting, look for a derangement in the control valve locking device or a relief valve which is stuck in a closed position.

Allow the control rod to set in the neutral position. If the implement drops slowly, it is an indication the check valves leak. A leaky power lift cylinder will also cause the implement section to drop slowly and may be checked by inspecting the piston rod guide for leakage. Check all pipe and hose connections.

PUMP

512. An exploded view of a typical hydraulic pump is shown in Fig. IH 490. Other than installing a seal ring and gasket package, there is little actual repair work which can be accomplished on the pump.

To remove the pump, first crank engine until number one piston is coming up on compression stroke and continue cranking until first notch on crankshaft pulley is in register with pointer on crankcase front cover.

Remove the ignition unit and the hydraulic lines flange from pump. Cover the openings in the hydraulic lines to prevent the entrance of dirt and remove pump from engine. With pump openings covered thoroughly wash pump in a suitable solvent to remove any accumulation of dirt. Remove nut retaining drive gear to drive shaft and using a suitable puller as shown in Fig. IH491, remove the drive gear and drive gear Woodruff key. Remove cap screws retaining cover to pump body, bump pump drive shaft on a wood block to loosen cover from pump body and remove cover. Remove seal rings (4—Fig. IH490), fiber washers (3), spring (B) and ring gasket (5). Press drive shaft seal (2) out of pump cover. Tap drive shaft (F) on a wood block to loosen bearings (C & D) from pump body, then remove the

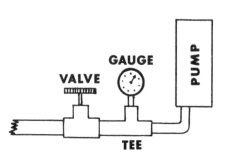

Fig. IH490A—Cut-off valve and gage used to check the hydraulic system pressure.

Fig. IH490—Exploded view of a typical engine driven hydraulic pump. Numbered items show contents of the seal and gasket repair package.

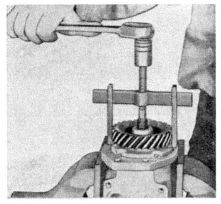

Fig. IH491—Removing drive gear from hydraulic pump. Care should be taken to avoid damaging the gear during this operation.

2. Seal	6. Pin seal (some
3. Fiber washer	models)
4. Seal rings	7. "O" rings

A. Cover	E. Driven gear
B. Springs	F. Drive gear
C. & D. Bearings	I. Body

bearings. Pin seal (6), used on some models, will come out with the bearings.

Remove the ignition unit drive lug (J) and remove gears (E & F). Tap body on wood block to remove bearings (G & H). Identify the bearings so they can be installed in their original position. Press seal (2) out of pump body.

Clean all metal parts in a suitable solvent and dry them with compressed air. If seal contacting surfaces on drive gear shaft are not perfectly smooth, polish them with fine crocus cloth and rewash the drive gear and shaft.

When reassembling, lubricate all parts with clean IH Touch-Control Fluid and use new gaskets and seals.

Install drive shaft seal (2) in body with lip toward center of pump. Install bearings (G and H) in their original position with milled slot on pressure side. Install gear (E) with tool marked side of gear toward pump body. Install gear (F), being careful not to damage seal (2) in body. A

seal jumper (Fig. IH492) can be used to avoid damaging the seal. Install bearings (C and D—Fig. IH490) and on models so equipped, install pin seal (6). Notice that long bearing (D) fits over the drive shaft. Install seal ring (5) and the ten springs (B). Install seal (2) in cover so that lip of seal faces center of pump. Install back up washer (3) and seal ring (4) in cover. Install pump cover, using the seal jumper to avoid damaging the seal. Install cover cap screws and tighten them to a torque of 25 Ft.-Lbs. Install the ignition unit drive lug, drive shaft Woodruff key and drive gear.

When installing the pump, reverse the removal procedure and check the ignition timing.

CONTROL VALVES & RESERVOIR
Series Super MTA

513. OVERHAUL. Before disassembling the reservoir and control valves, thoroughly wash the unit to remove any accumulation of grease or dirt.

Remove cap screws retaining back plate to reservoir, move the back plate assembly until cam lever (8—Fig. IH493) is free from inner control lever (26) and lift the back plate assembly from reservoir. Remove locking clip (13) and withdraw spring (3). Remove the relief valve cap. Remove pin (12) and remove the cam and piston valve lever (8) and spring (11).

Unscrew the relief valve body (20) and remove the relief valve spring (19), spring sleeve (18), ball (17), ball seat (16) and gasket (15). Unbolt and remove cover (42), identify the differential check valve pin (7) and the hold-up ball pin (44) with respect to their bores so they can be reinstalled in the same position and remove the pins (7 & 44).

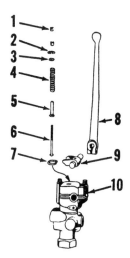

Fig. IH494—Exploded view of the auxiliary control valve used on the MTA and W6TA series tractors.

1. Control valve centering spring spacer	5. Centering spring sleeve
2. Centering spring lock	6. Centering spring rod
3. Washer	7. Seal ring
4. Centering spring	9. Yoke
	10. Body

Surfaces must be smooth

Fig. IH492—Home made tool which can be used to avoid damaging hydraulic pump seals when installing the pump drive shaft. Tool can be made from 7/8 inch diameter steel.

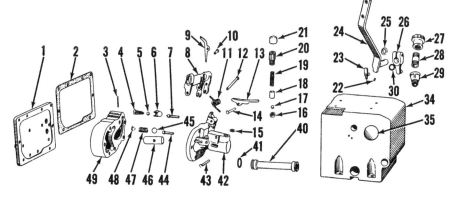

Fig. IH493—Exploded view of the Super MTA series hydraulic system reservoir and control valves.

1. Back plate	9. Cam plate	34. Reservoir
2. Gasket	10. Pin	35. Expansion plug
3. Locking clip spring	11. Lever spring	40. High pressure passage tube
4. Differential check valve spring	12. Lever pin	41. Seal ring
5. Differential check valve ball	13. Locking clip	42. Valve body cover
6. Differential check valve	14. Pin	43. Pin
7. Differential check valve lift pin	15. Relief valve gasket	44. Hold-up ball pin
8. Cam and piston valve lever	16. Ball pressure seat	45. Hold-up ball
	17. Ball	46. Piston
	18. Relief valve spring sleeve	47. Hold-up ball spring
	19. Relief valve spring	48. Hold-up ball stop
	20. Relief valve body	49. Valve body
	21. Relief valve cap	
	22. Roll pin	
	23. Control lever latch	
	24. Control lever and shaft	
	25. Key	
	26. Inner lever	
	27. Breather cap	
	28. Nipple	
	29. Elbow	
	30. Seal ring	

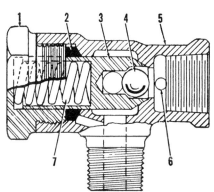

Fig. IH495 — Sectional view of the Super MTA series delayed lift valve.

1. Valve plug	4. Ball
2. Rubber seal	5. Valve housing
3. Valve piston	6. Pin
	7. Spring

Remove body (49) and withdraw differential check valve (6), balls (5 & 45), springs (4 & 47) and stop (48).

Wash and inspect all internal parts and renew or recondition any that show damage or wear.

When reassembling, reverse the disassembly procedure and make certain that spring (11) is installed so that it forces cam plate down against pins (7 & 44). When installing the back plate, be sure that high pressure tube (40) is in position and tip the back plate to engage the cam lever (8) with the inner control lever (26).

AUXILIARY CONTROL VALVE

514. OVERHAUL. An exploded view of the auxiliary control valve is shown in Fig. IH494. The disassembly and reassembly procedure is evident after an examination of the unit. Normal overhaul consists of disassembling, cleaning and renewing any damaged parts.

Fig. IH496—Exploded view of the Super MTA series selective lift control.
1. Adjustment valve lever
2. Washer
3. Oil seal
4. Groove pin
5. Adjustment valve
6. Valve housing manifold
8. Selector lever
15. Cap screw
16. Manifold ball stop
17. Ball 5/8" diam.
18. Ball 7/16" diam.
19. Valve shaft
20. Shaft bearing
21. Snap ring
22. Valve lever
25. Rivet
26. Selector lever pivot block
27. Groov pin
28. Spring
29. Gasket
30. Valve housing
31. Ball stop

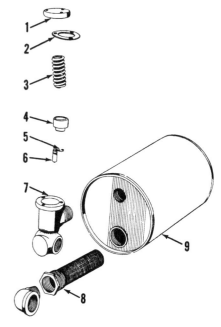

Fig. 497—Exploded view of the Super W6TA hydraulic system reservoir.

1. Safety valve cover	5. Safety valve piston ring
2. Cover gasket	6. Piston
3. Safety valve spring	7. Safety valve body
4. Safety valve sleeve	8. Oil strainer
	9. Reservoir tank

HYDRAULIC LIFT-ALLS
(SERIES H-M EXCEPT MTA)

The Lift-All pumps of early manufacture were set to kick off at 450-500 pounds pressure in the system. Later pumps are set to kick off at 750-800 pounds pressure. These higher pressure units can be identified by a yellow figure "8" painted on the outside of the reservoir.

An improvement package (15261-E) is available to change low pressure pumps to high pressure pumps. Failure of the improved pump to build a pressure of 750-800 pounds may be caused by too great a clearance between gears and pump cover. Maximum clearance is 0.005. The low pressure pump maintained a clearance of 0.007. Where the increased clearance is found, it may be either necessary to grind down the pump cover to decrease the clearance or renew the pump parts.

Oil capacity for the Lift-All is 6 quarts.

TROUBLE SHOOTING

519. PUMP UNIT. Before removing a faulty Lift-All from a tractor, it is advisable to first make a test to determine the cause of failure.

Disconnect the hose from the outlet elbow at the pump unit and install a ½ inch tee between the elbow and hose fitting as shown in Fig. IH499. Install a hydraulic pressure gauge, SE-1338 or equivalent, in the tee.

Start tractor engine and operate the pump unit to warm the oil. Throttle the engine to idling speed and pull the Lift-All control rod rearward. Note the oil pressure reading which should be 750 pounds or higher. If the pressure is 750 pounds or higher, the pump is in good condition. If the pressure is lower, it may be caused by any one of the following:
(A) Oil low in reservoir.
(B) Excessive clearance in pump.
(C) Leaky relief valve.
(D) Leak in reservoir (high pressure passages).
(E) Leak past bushings on drive shaft or oil seal.
(F) Slot in back plate too long, causing oil to by-pass pump gears.

Move the Lift-All control rod forward. Speed up the engine and again pull the control rod rearward, but this time release the rod. If the rod does not remain in the lifting position, it is an indication that the locking device is not engaging properly. In some cases the relief valve may be sticking in its expanded position.

With the control lever pulled rearward and upon completing the lifting operation the control lever should move automatically to the neutral position. Gauge reading should be 750-800 pounds. If the reading is less than 750 pounds, the cause may be:
(A) Leaky relief valve.
(B) Wrong relief valve spring.
(C) Relief cap missing.

Allow the control rod to set in the neutral position. If the implements drop slowly, it is an indication the check valves leak. A leaky power lift cylinder will also cause the implement section to drop slowly and may be checked by inspecting the piston rod guide for leakage. Check all pipe and hose connections.

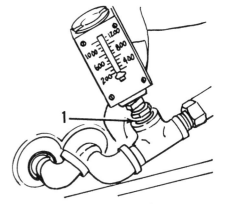

Fig. IH 499—Series H and M pump pressure gage installation. Gage (IH No. SE-1338) is installed in reducing sleeve (1).

R&R PUMP UNIT

520. To remove the pump unit which is located on the underside of the clutch housing, proceed as follows: Disconnect control lever and hoses. Unscrew pipe elbows from both sides of pump unit. Remove filler pipe and bayonet gauge. Remove dust pan from lower side of clutch housing.

Disconnect and remove coupling and sleeves. See Fig. IH500. A pair of "T" wrenches similar to those shown in Fig. IH501 will make the removal of the unit much easier.

Insert the "T" wrenches, as shown in Fig. IH502, into the offset holes located on lower side of pump unit. Remove the four cap screws retaining pump unit to clutch housing and remove pump.

To reinstall the pump unit, reverse the removal procedure, using extreme caution to prevent dirt from entering pump lines and fittings.

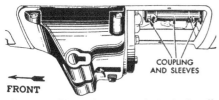

Fig. IH 500—Series H and M hydraulic pump installation, with clutch housing dust pan removed.

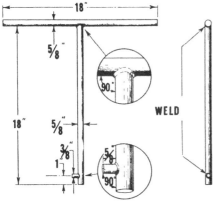

Fig. IH 501—Details of special tool which is used for removal and reinstallation of series H and M pump unit.

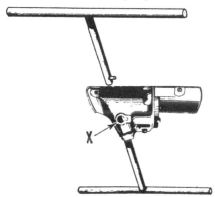

Fig. IH 502—Inserting special "T" tools into offset holes at point (X) for pump removal on series H and M.

DISASSEMBLY AND REASSEMBLY

521. To remove pump from oil reservoir remove the ten retaining cap screws. Break pump loose from the gasket. Move pump until the inner control lever is released from the cam and piston valve lever and remove the pump.

Turn the pump so that the oil intake faces up. Place a flat piece of steel (1/8 x 3/4 x 6) into the oil intake opening to prevent the gears from rotating while removing pump coupling from the shaft, by turning coupling in a counter-clockwise direction, as shown in Fig. IH503.

Remove cotter key and pin attaching locking lever to pump cover and remove locking lever and spring. Remove relief valve cap. Remove cotter key and pin attaching cam and piston valve lever to pump cover and lift off the lever and spring, as shown in Fig. IH504.

Unscrew spring retainer from pump cover and remove relief valve spring, sleeve and ball. Using a small tool as

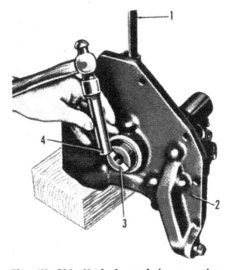

Fig. IH 503—Method used in removing coupling from series H and M pump shaft.
1. 1/8 x 3/4 x 6 steel bar   3. Coupling
2. Pump unit   4. Brass rod

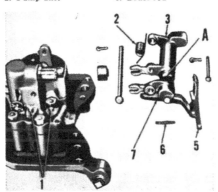

Fig. IH 504—Series H and M cam and piston valve lever removed from pump. Notice the relief at point (A).
1. Ball lift pins   5. Locking lever
2. Lever spring   6. Locking lever spring
3. Valve lever   7. Cam

shown in Fig. IH505, remove the ball seat.

Remove pump cover as shown in Fig. IH506, and remove pump gears and check valves as shown in Fig. IH 507.

Reassemble by reversing the disassembly procedure.

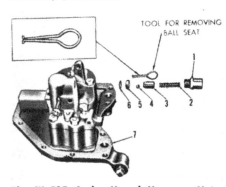

Fig. IH 505—Series H and M pump. Note special tool for removing relief valve parts.
1. Spring retainer   5. Ball seat
2. Relief valve spring   6. Ball seat
3. Spring sleeve   gasket
4. Relief valve ball   7. Pump unit

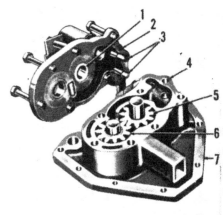

Fig. IH 506—Series H and M pump with cover removed
1. Drive shaft bushing   5. Drive gear
2. Pump cover   6. Driven gear
3. Ball lift pins   7. Pump back
4. Pump body   plate

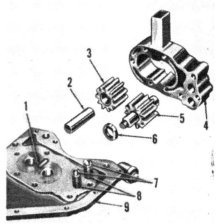

Fig. IH 507—Series H and M pump with pump body and gears removed.
1. Relief slot   6. Oil seal
2. Driven gear shaft   7. Check valves
3. Driven gear   8. Check valve
4. Pump body   springs
5. Drive gear   9. Pump back plate

OVERHAUL

522. Inspect for gear wear on pump cover, back plate and inside of pump body. Manufacturing dimensions are as follows:

Pump body thickness....2.001-2.003
Gear thickness 1.998-2.000
Allowable clearance 0.001-0.005
*Clearance between pump body
and gears 0.003-0.007
*If clearance exceeds 0.010, renew pump body.

Inspect drive shaft bushings and also driven gear bushing. Inside diameter of these bushings is 1.001-1.002. New bushings are pre-sized and should be pressed into the gears until flush with the end of the gear. This will leave a small space between the bushings.

Inspect the two check valve seats in the pump body and the two check valve balls for rust or damaged spots. Valve seats may be lapped in by using a steel ball 1/32 inch larger than the valve ball and fine grinding compound. Finish the lapping operation, using a paste made from Bon-Ami and water.

Make sure bearing retainer is bent so that it will hold the drive shaft bushing in the pump cover flush with the outside face of the cover as shown in Fig. IH508.

Inspect relief valve ball and seat for rust or damaged spots.

Check relief valve springs. Relief valve spring free length 1 5/16 ± 1/32 inch. Check valve spring free length 1⅛ ± 1/32 inch.

Inspect cam and piston valve lever for being the latest which has a relief at "A" as shown in Fig. IH504.

Inspect oil seal. New oil seal should be installed in the back plate with the lip facing in.

Inspect oil reservoir in model M lift for leaks.

POWER CYLINDERS

526. Power cylinders have been furnished in three different sizes: 1¾, 2¼ and 3 inch inside diameter of cylinder.

The 1¾ inch power cylinder has been furnished with two types of piston construction, Fig. IH510, IH511 or IH512.

The earlier design of the 2¼ inch power cylinder employed a hollow piston with a check valve in the pressure end, and another in the opposite end which permitted the oil to return to the pressure end from the opposite end on the return stroke of the piston. The check valves are eliminated on the later models, Fig. IH 512A.

All 3 inch power cylinders have two check valves in the piston which permits the oil to return to the pressure side of the piston on the return stroke, Fig. IH513.

DELAYED POWER LIFT VALVE

527. The delayed lift valve, Fig. IH514, is used to delay the lifting of the rear section of a cultivator until the front section of same comes out of the ground. It is installed between cylinder hose and power cylinder used for the rear cultivator section.

A pressure of 425-475 pounds is required to lift the valve off its seat. Earlier delayed lifts, as used with the 500 pound Lift-Alls, were set to open at 275-300 pounds. These valves can be set to open at 425 to 475 pounds by installing a ⅜ washer behind the spring in the plug.

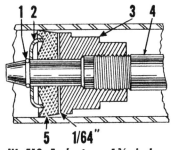

Fig. IH 508—Series H and M pump. Arrow shows installation of bushing retainer.

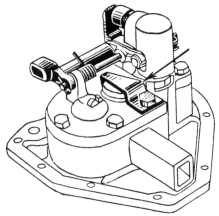

Fig. IH 510—Early type 1¾ inch power cylinder correctly assembled. Note 1/64 inch clearance between cup and piston nut.

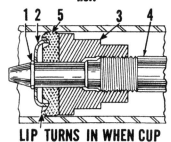

LIP TURNS IN WHEN CUP IS COMPRESSED

Fig. IH 511—Early type 1¾ inch power cylinder incorrectly assembled.

1. Snap ring	4. Piston
2. Retainer	rod
3. Piston nut	5. Cup

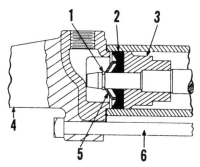

Fig. IH 512—Later type 1¾ inch power cylinder.

1. Snap ring	4. Cylinder head
2. Rubber cap	5. Cup retainer
3. Piston	6. Bolt

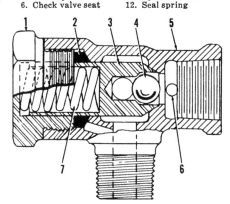

Fig. IH 512A—A 2¼ inch power cylinder.

1. Piston nut	4. Piston
2. Gasket	cup
3. Spring	5. Piston

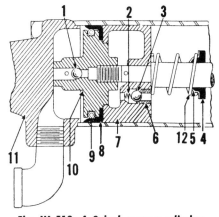

Fig. IH 513—A 3 inch power cylinder.

1. Check ball	7. Piston
2. Check ball spring	8. Piston cup
3. 5/16" check ball	9. Expansion spring
4. Floating seal	10. Piston cup nut
5. Seal cup	11. Cylinder head
6. Check valve seat	12. Seal spring

Fig. IH 514—Series H and M delayed lift valve assembly.

1. Valve plug	4. Ball
2. Rubber seal	5. Valve housing
3. Valve piston	6. Pin

DROP-RETARDING VALVE

528. The drop-retarding valve restricts the return of oil from a power cylinder when an implement is lowered. Always install the retarding valve, Fig. IH515, on the end of the high pressure hose to be connected to the elbow on the pump unit. If the valve is inserted on the power lift cylinder, it will retard the lift instead of the drop.

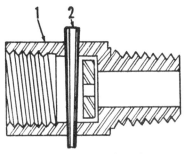

Fig. IH 515—Series H and M drop retarding valve assembly.
1. Valve body   2. Tapered pin

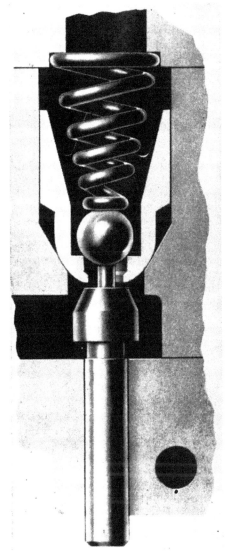

Fig. IH 516—Super M and W6 series oil return valve. The valve permits either fast or slow return of oil to the reservoir.

HYDRAULIC "TOUCH CONTROL" (SERIES A-C-Cub)

The hydraulic power lift ("Touch Control") system is composed of three basic units: The gear type pump, which is gear driven from the timing gear train; a single or double work cylinder and valves unit which is mounted on the torque tube; and one or two rockshaft assemblies bolted to the forward end of the valves unit.

The system as used on the model C has two work cylinders, twin rockshafts and four rockshaft power arms. The Super A, Super AV and Super C systems are similar to the Model C, except that three rockshaft power arms are used.

The "Touch Control" system which is used on the Cub is similar in operation to the above models except that the system is smaller and a single rockshaft and one rockshaft operating (work) cylinder is used.

Note: *The maintenance of absolute cleanliness of all parts is of utmost importance in the operation and servicing of the hydraulic system. Of equal importance is the avoidance of nicks or burrs on any of the working parts.*

LUBRICATION AND BLEEDING

535. To refill the reservoir and bleed the system, after the system has been drained, proceed as follows: Refill the reservoir with IH "Touch Control" Fluid. Start the engine and run at approximately 650 rpm. With the filler plug removed, move the control levers back and forth 10-12 times; then, place the levers in the rear position and stop the engine. Add sufficient fluid to bring the reservoir fluid level to within ½ inch of the filler opening. Capacity of the complete system is approximately 8¼ pints for series A and C and 4¼ pints for the Cub.

CAUTION: If the system is to be flushed, do not use kerosene. It is recommended that IH "Touch Control" Fluid be used.

TROUBLE SHOOTING

536. If the "Touch Control" system does not operate properly, it is advisable to apply a few quick checks to determine which unit is at fault. To check the system, install a Schrader (IH No. SE-1338-A) gage or equivalent in the position shown in Fig. IH520 for Series A and C or in the upper port of the rear manifold flange for the Cub. Load the system with two rear wheel weights attached to the rear rockshaft arm of a rear mounted implement rockshaft. Start the engine, operate the lift system and refer to the trouble diagnosis chart which follows:

Note: Keep in mind that the gage should show high pressure (1100-1500 psi) only during the movement of rockshaft arms. When the rockshaft arms have completed their travel, the system will return to low pressure (15-40 psi). The fact that this low range of pressure will not be indicated on the SE-1338-A gage is of no importance, since the low range of pressure is not a factor in the trouble shooting procedure.

A. System Will Not Lift Load, High Gage Pressure.
    1. Binding or scored rockshaft bearings.
    2. Damaged implement.
    3. Defective cylinder head gasket.
B. System Will Not Lift Load, Low Gage Pressure.
    1. No oil in system—check for leaks.
    2. Faulty pump.

Fig. IH 520 — Schrader (IH No. SE-1338-A) gage installed on series A and C. Gage installation on Cub is similar, except the gage is installed in the upper port of the flange.

3. Regulator and/or safety valves stuck.
4. Weak or broken safety valve spring.
5. Excessive safety valve clearance in its bore.

C. Lift Cycle Slow (More than 2 seconds required to lift load), Low Gage Pressure.
  1. Same as 2, 3, 4 and 5 under condition B.
  2. Orifice plug opening too large (more than 0.030).
  3. Internal pipe plugs (clean-out plugs) loose or missing.

D. Gage Shows High Pressure When Control Levers Are Stopped At Either Of The Extreme Positions, Low Pressure When Levers Are Stopped At Intermediate Positions.
  1. Faulty implement. Implement preventing full rockshaft travel.
  2. Stop clips out of adjustment.

E. Gage Shows High Pressure With Control Levers And Rockshaft In Any Position.
  1. Orifice plug stopped up.
  2. Stuck regulator valve.
  3. Internal pipe plugs (clean-out plugs) loose or missing.
  4. Cracked or faulty work cylinder block.

F. Load Oscillates When Engine Is Running, Drops Slowly When Engine Is Stopped.
  1. Oil leaking past work cylinder piston due to:
    a. Check valves not seating in bushings.
    b. Defective seal rings on check valve bushings (Not Cub).
    c. Defective work cylinder piston seal rings.
    d. Defective cylinder head gasket.
    e. Cracked or faulty work cylinder block or head.

G. Same As Condition F, Except Load Stays In Raised Position.
  1. Refer to condition F.
  2. Work cylinder piston inner seal ring leaking.
  3. Leak at weld between piston head and sleeve.
  4. Thermal relief valves leaking.

H. Loss Of Oil from System, No External Leaks.
  1. Pump drive shaft seal leaking oil into crankcase.

I. **Control Levers** Creep When Rockshaft Is In Motion.
  1. Insufficient friction at control levers.
  2. Sprung or bent control rods.

3. Binding control spool—free up walking beam.

J. Gage Pressure Too High (More than 1500 psi).
  1. Stuck or binding safety valve.
  2. Faulty safety valve spring (free length, 1 15/16 inches; should be 1¼ inches long under 61-67 lbs.).

ADJUSTMENT

537. STOP CLIPS SERIES A-C. To adjust the "Touch Control" stop clips on series A and C, proceed as follows: Install a Schrader (IH No. SE-1338A) gage or equivalent in place of the ¼ inch Allen head pipe plug in the pump output side of the manifold rear flange as shown in Fig. IH520. Place the left hand control lever and its front stop about mid-position of the quadrant, tighten the stop thumb screw and wire the control lever to the stop. Start engine and run at half speed. Slip the right front quadrant stop past the right hand control lever and move the control lever fully forward. At this time, when the rockshaft has completed its stroke, the system should remain on high pressure (1100-1500 psi) as shown on the gage. If the system does not remain on high pressure, adjust the length of the right hand control rod until proper condition is obtained. Now, watching the gage, slowly move the right hand control lever toward rear until system returns to low pressure but not far enough to move the rockshaft from its extreme position. Then, scribe a line across the top edge of both rockshaft arm shields. Now, again move the right hand control lever toward rear until the outer rockshaft arm has moved back approximately ½ inch as shown by the distance between the scribed lines on the rockshaft arm shields.

Without moving the rockshaft arms from the previously mentioned position, move the right hand stop clip

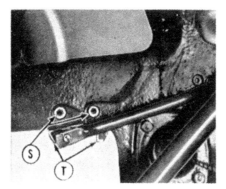

Fig. IH 521—Adjust "Touch Control" stop clip (T) with cap screws (S).

(T—Fig. IH521) forward until the clip firmly contacts the control valve operating lever pin and lock the clip in this position with the two Allen head cap screws (S).

Operate the right hand control lever back and forth several times and check the extreme forward position of the rockshaft to make certain that the scribe lines on the arm shields stop ½ inch apart. Note: It may be necessary to readjust the clip slightly to maintain the ½ inch dimension. Paint over the scribe lines so as not to confuse with new lines made when adjusting the left stop.

Procedure for adjustment of the left hand stop clip is the same as for the right, except the right hand control lever must be wired in its mid-position on the quadrant.

538. STOP CLIP CUB. To adjust the "Touch Control" stop clip on the Cub, proceed as follows: Install a Schrader (IH No. SE-1338-A) gage or equivalent in the upper port of the rear manifold flange. Loosen the stop clip retaining cap screws. Start engine and run at half speed. Move the control lever fully forward. At this time, when the rockshaft has completed its stroke, the system should remain on high pressure (1100-1500 psi) as shown on the gage. If the system does not remain on high pressure, adjust the length of the control rod until the proper condition is obtained. Now, watching the gage, slowly move the control lever toward rear until the system returns to low pressure, but not far enough to move the rockshaft from its extreme position. Then, with the rockshaft in the extreme forward position, measure the distance between the pin in the rockshaft arm and the carburetor bowl cover and remember the measurement. With rule in same position, move the control lever rearward until rockshaft has moved rearward ⅜ inch.

Without moving the rockshaft arms from the previously mentioned position, move the stop clip forward against the yoke pin and tighten the clip retaining cap screws.

Operate the control lever several times and check to make certain that the ⅜ inch differential is maintained. Note: It may be necessary to readjust the clip slightly to maintain the ⅜ inch differential.

PUMP UNIT

539. REMOVE AND REINSTALL. To remove the gear type hydraulic pump, which is gear driven from the camshaft gear, drain the hydraulic system and remove the hydraulic lines (manifold). Remove the pump attaching cap screws and lift pump from tractor.

Install the hydraulic pump by reversing the removal procedure and bleed the "Touch Control" system as outlined in paragraph 535.

540. OVERHAUL. See Fig. IH522. Overhaul of the hydraulic pump is limited to disassembling, cleaning and installing a gasket and seals package.

If parts, other than gaskets and seals are excessively worn or damaged, it will be necessary to renew the complete pump unit which is available from the International Harvester Co. on an exchange basis.

To renew the pump gaskets and seals, proceed as follows: Mark the pump body and cover so they can be reassembled in the same relative position and remove the cover. Mark the exposed end of the driven (idler)

gear so it can be installed in the same position and remove the drive and driven gears. Identify the bearings with respect to the pump body and cover so they can be reinstalled in the same position and remove the bearings. The procedure for further disassembly is evident after an examination of the unit.

Check the pump parts against the values given below. If any of the parts are worn in excess of the values listed, the pump should be exchanged.

Clearance between body bore and
gear 0.0005-0.004
Clearance between shaft journals
and bearing bores...... 0.0015-0.005
Gear thickness (Series A &
C) 0.420-0.425
Gear thickness (Cub).... 0.434-0.4385
Body bearing flange thickness
(Series A & C)........ 0.180-0.187
Body bearing flange thickness
(Cub) 0.150-0.156
Cover bearing flange thickness
(Series A & C) 0.370-0.3755
Cover bearing flange thickness
(Cub) 0.365-0.380

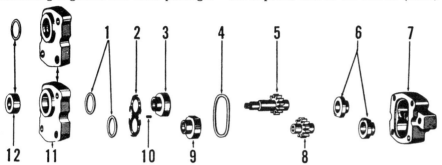

Fig. IH 522—Exploded view of "Touch Control" pump. Items (11) and (12) were used on early production pumps only; the parts shown above items (11 & 12) are typical of the later construction.

1. Cover seal rings	4. Body seal ring	8. Driven gear
2. Cover bearing spring	5. Drive gear	10. Bearing seal pin
3 & 9. Cover bearings	6. Body bearings	11. Pump cover
	7. Pump body	**12. Drive shaft seal**

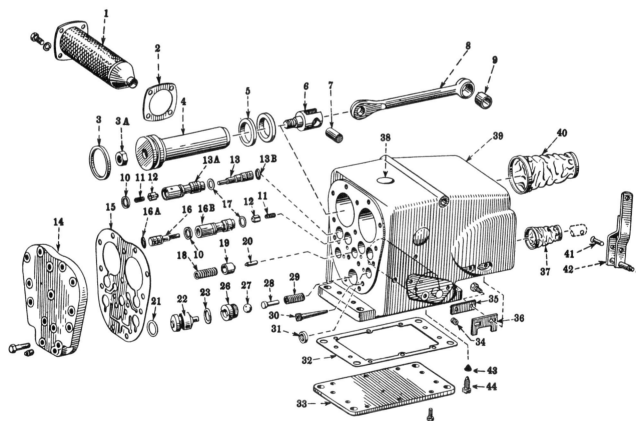

Fig. IH 523—Exploded view of series A and C—hydraulic cylinder and valves unit.

1. Strainer	9. Rod bushing	15. Head gasket	21. Pressure regulator valve piston seal ring
2. Gasket	10. Check valve center bushing seal ring	16. Valve actuator	
3. Piston head seal	11. Check valve spring	16A. Actuator seal ring	22. Pressure regulator valve piston
3A. Connecting rod nut	12. Check valve	16B. Check valve front bushing	23. Pressure regulator valve seat seal ring
4. Piston	13. Valve actuator	17. Check valve bushing center seal ring	26. Pressure regulator valve seat
5. Piston sleeve seal	13A. Check valve rear bushing	18. Safety valve spring	27. Ball
6. Yoke	13B. Actuator seal ring	19. Safety valve sleeve	
7. Yoke pin	14. Cylinder head	20. Safety valve piston	
8. Connecting rod			

28. Ball rider	36. Stop clip	
29. Rider spring	37. Control valve boot	
30. Orifice plug & screen	38. Expansion plug	
31. Control valve seal ring	39. Cylinder block	
32. Gasket	40. Piston sleeve boot	
33. Cover	41. Control valve pin	
34. Plug	42. Operating lever	
35. Stop clip block	43. Relief valve screen	
	44. Relief valve	

Clearance between bearing journals and recesses in body and cover 0.001-0.006

Diametral clearance between bearing flange and flange recess in cover (Series A & C) ..0.0015-0.006

Diametral clearance between bearing flange and flange recess in cover (Cub) 0.0009-0.006

Note: Small nicks and/or scratches can be removed from the body cover, shafts and gears, and bearings by using crocus cloth or an oil stone, providing the limits given above are not exceeded. When dressing the bearing flanges, however, make certain that the flange thickness of both bearings in either pair are identical.

Lubricate all pump parts with IH "Touch Control" Fluid prior to reassembly. When bearings are installed in the pump body and cover, make certain that the clearance between the bearing flats does not exceed 0.0005. On early production pumps (pumps equipped with a lip-type drive shaft seal) use caution when installing the shaft through the seal to avoid damaging the seal.

CYLINDER AND VALVES UNIT

542. **REMOVE AND REINSTALL.** To remove the hydraulic cylinder and valves unit, drain the system and remove the hydraulic lines (manifold). Remove the hood and fuel tank and disconnect wires and control rods from the cylinder. On models so equipped, remove the heat indicator sending unit from the cylinder block. Remove the cap screws retaining the unit to the torque tube and lift the unit from the tractor.

Note: On some models, the "Touch Control" heat indicator sending unit was soldered to the cylinder block strainer. When such cases are encountered, it is advisable to remove the sending unit and strainer, as an assembly, from the cylinder block.

Caution: If for any reason the tractor must be started after the hydraulic cylinder block is removed, the hydraulic pump must be removed. The pump will fail quickly when it is disconnected from the oil supply.

543. **OVERHAUL.** Overhaul of the "Touch Control" cylinder and valves block is limited to completely disassembling the unit, cleaning and renewing any damaged parts. Refer to Fig. IH523 or 525. To disassemble the unit, proceed as follows: Remove the cylinder head, bottom inspection plate and oil strainer. On models so equipped, remove the top inspection plate. Note the position of all parts so they can be reinstalled in their same position and remove check valves, screw plug, pressure regulator piston, etc., from the block. Turn the block around and remove the rock-shaft assembly, dust boots, control valves and pistons. The balance of the disassembly procedure is evident after an examination of the unit.

When reassembling the unit, dip all parts in clean "Touch Control" fluid and lubricate all "O" ring seals with Vaseline or equivalent. When assembling the piston and connecting rod, tighten the rod yoke retaining nut to a torque of 110 ft.-lbs. Tighten the cylinder head retaining cap screws to a torque of 45 ft.-lbs.

Reinstall the unit on the tractor, bleed the system as outlined in paragraph 535 and adjust the unit as in paragraph 537.

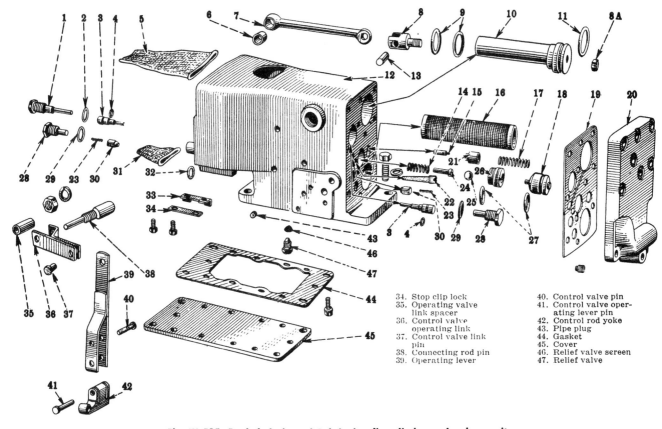

34. Stop clip lock	40. Control valve pin
35. Operating valve link spacer	41. Control valve operating lever pin
36. Control valve operating link	42. Control rod yoke
37. Control valve link pin	43. Pipe plug
38. Connecting rod pin	44. Gasket
39. Operating lever	45. Cover
	46. Relief valve screen
	47. Relief valve

Fig. IH 525—Exploded view of Cub hydraulic cylinder and valves unit.

1. Check valve actuator lower stop	7. Connecting rod	13. Yoke pin	21. Safety valve sleeve
2. Stop washer	8. Yoke	14. Rider spring	22. Orifice plug & screen
3. Check valve actuator	8A. Yoke nut	15. Safety valve piston	23. Check valve spring
4. Actuator seal ring	9. Piston sleeve seal ring	16. Strainer	24. Valve ball rider
5. Piston sleeve boot	10. Piston	17. Safety valve spring	25. Ball
6. Connecting rod bushing	11. Piston head seal ring	18. Pressure regulator valve piston	26. Regulator valve seat
	12. Cylinder block	19. Head gasket	27. Regulator valve piston seal ring
		20. Cylinder head	
28. Check valve actuator plug			
29. Actuator plug washer			
30. Check valve			
31. Control valve boot			
32. Control valve seal ring			
33. Stop clip			